Springer Aerospace Technology

The series explores the technology and the science related to the aircraft and spacecraft including concept, design, assembly, control and maintenance. The topics cover aircraft, missiles, space vehicles, aircraft engines and propulsion units. The volumes of the series present the fundamentals, the applications and the advances in all the fields related to aerospace engineering, including:

- structural analysis,
- aerodynamics,
- aeroelasticity,
- aeroacoustics,
- flight mechanics and dynamics
- orbital maneuvers,
- avionics,
- systems design,
- materials technology,
- launch technology,
- payload and satellite technology,
- space industry, medicine and biology.

The series' scope includes monographs, professional books, advanced textbooks, as well as selected contributions from specialized conferences and workshops.

The volumes of the series are single-blind peer-reviewed.

To submit a proposal or request further information, please contact: Mr. Pierpaolo Riva at pierpaolo.riva@springer.com (Europe and Americas) Mr. Mengchu Huang at mengchu.huang@springer.com (China)

The series is indexed in Scopus and Compendex

Paolo Teofilatto

Trajectories of Launch and Reentry Systems

Springer

Paolo Teofilatto
School of Aerospace Engineering
Sapienza University of Rome
Rome, Italy

ISSN 1869-1730 ISSN 1869-1749 (electronic)
Springer Aerospace Technology
ISBN 978-3-031-75759-4 ISBN 978-3-031-75760-0 (eBook)
https://doi.org/10.1007/978-3-031-75760-0

This Springer imprint is published by the registered company Springer Nature Switzerland AG
The registered company address is: Gewerbestrasse 11, 6330 Cham, Switzerland

This book is dedicated to the memory of Antonio and Tarcisio Teofilatto who supported and encouraged me along this path of research

Preface

In 1995 I was assigned to the Flight Mechanics course at the School of Aerospace Engineering of Rome. My Ph.D. supervisor in Aerospace Engineering, Prof. Filippo Graziani, asked me to change the programme of the course from a classical aeronautical course to the flight mechanics of launch and re-entry systems. At national level, this was the first academic course on these subjects, for which there were very few references in the literature. However, the School had operated the Scout space launch vehicle from the San Marco platform for 20 years (1967–1988), so I had a heritage to work from. In the last 30 years, the content of the course has been extended and deepened, also by working as a consultant for some aerospace companies such as Avio, ELV and MBDA. Important improvements came from the comments and feedback from the students who have attended the course.

The book takes advantage of several discussions, hints and collaborative work had with Stefano Carletta, Antonio Rinalducci, Mauro Balduccini, Alessandro Gabrielli, Emanuele di Sotto and Emilio De Pasquale.

Some of the developments in optimal staging and optimal exo-atmospheric trajectories could not have taken place in their present form without the support of Mauro Pontani. He was a former Ph.D. student of mine, but over time I have learned more from him than he has from me.

I am also deeply indebted to Bill Harvey for instilling in me his love of mathematics during my Ph.D. at King's College London under his supervision. Although I have moved away from pure mathematics, I still feel his influence and guidance. In this vein, the book proposes some original analytical formulae useful in the design of launch and re-entry trajectories.

Various topics that are in general limited to internal industry and agency reports are presented here: this is due to the heritage of the Scout launch vehicle held at the School. Although long since decommissioned, the Scout format is representative of the new generation of launchers designed to serve the microsatellite market, and the

San Marco sea platform is representative of the new options that are being considered for autonomous access to space for those countries that do not have a suitable geographical location.

Rome, Italy
September 2024

Paolo Teofilatto

Contents

Chapter 1
Launch Systems

Abstract This chapter offers a broad and in-depth description of all the subsystems of a space launcher, with a high level of detail thanks to the legacy of the Scout launcher at the School of Aerospace Engineering in Rome. Although retired in 1994, Scout is representative of the format of the new generation of launchers designed for the microsatellite market. It used all the control systems that can be accommodated on a launcher: aerodynamic control, thrust vector control, gas jet control and spin stabilisation. Scout was launched from the San Marco sea platform in Malindi (Kenya), owned by the School of Aerospace Engineering of Rome. The launch activity in Malindi started in 1967 and ended in 1988 without a failure. Today, the sea platform is one of the new options for autonomous access to space for those countries that do not have a suitable geographical position. A detailed description of the infrastructures and operations of the launch base is given in the chapter. The chapter is divided into nine sections, including optimal staging, description of stage separation systems (hot and cold), propulsion system, safe and arm, actuators and aerodynamics. Launch windows, launch azimuth selection and launch operations are described. Problems and solutions are presented.

1.1 Introduction

The task of a space launcher is to inject satellites into orbits. to do this, the launcher's engines have to provide the satellite with a considerable amount of energy. In addition to the engines, a launcher is made up of several subsystems:

- Stages: due to current technological limitations it is not possible to provide a spacecraft with the required variation of energy using a single engine: at least two engines are required so that the mass of the tank of the exhaust propellant can be released to reduce the rocket total weight. Space launchers have a number of stages between two and four.
- Propulsion system: is composed of propellant, tanks, combustion chamber and nozzle. The engines are fired and stopped by the commands given by the on board computer.

P. Teofilatto, *Trajectories of Launch and Reentry Systems*, Springer Aerospace Technology, https://doi.org/10.1007/978-3-031-75760-0_1

- Payload adapter and heat shields: The payload (spacecraft) that a space launcher can put into orbit is about the 1% of the total initial mass of the rocket. The payload has a mechanical interface with the rocket: the payload adapter. This device has a separation system to allow the spacecraft to be released at the orbit injection. During the atmospheric phase of the flight the payload is protected from the dynamical pressure loads and the heat flow by a covering structure, the heat shield.
- Avionics system: is the hardware related to the guidance, navigation and control of the launcher as well as the flight software loaded on the on board computer to implement the algorithms for the rocket maneuvers and to ensure dynamic stability.
- Safe and Arm: This system is responsible for the ignition of each stage and it is able to stop the flight by collapsing the launcher structure in case of severe malfunctioning detected by the on ground control station and/or by the on board system.
- Telecommunication and Electrical system: is the hardware and software needed to transmit the rocket telemetry during the flight, to receive commands from ground and to manage electrical power.

These systems will be reviewed in the following sections taking in consideration the actual case of the launcher Scout (Solid controlled orbital utility test system). This is a space launcher developed in 1957 at the NACA Langley Center and in service from 1961 to 1994. Scout was developed using "off the shelf" hardware, originally produced for military programs. The stages were developed using existing engines, in particular the first stage engine, called Algol, was a combination of the Jupiter Senior and the Navy Polaris; the second stage, called Castor, came from the Army MGM-29 Sergeant. The third (Antares) and fourth (Altair) stage engines were designed by Langley engineers adapting a version of the Navy Vanguard missile. The guidance and control was based on a pre-loaded attitude program with a time scheduling of the operations to be performed: although this was an open loop system, the reliability and accuracy of orbit injection were remarkable. Manoeuvres and control laws were actuated using all the possible kind of control systems that can be used on a launcher: aerodynamic control, thrust vector control, gas jet control, spin stabilization. Scout was launched from three sites: the launch base in Wallops Island (USA), the Vandemberg Space Center (USA) and the San Marco platform in Malindi (Kenya) owned by the School of Aerospace Engineering, of Rome. The launch activity in Malindi began on 1967 and ended in 1988 with no failure.

Although the project is more than 50 years old, with a total of 111 launches, Scout can be considered as a reference system for the new generation of small launchers being developed to meet to the market demand for the launch of nano and microatellites, targeting simple and low cost systems with a reduced launch time.

Scout launcher was composed by 4 stages, all having solid propellant [1]. The launcher total arrangement is shown in Fig. 1.1. The length of assembly of the four stages is 20 m and the total length is more than 22 meters, depending on the payload interface. The components of the launcher system are shown in Fig. 1.2, where also the interstages, heat shields and adapters are shown. Different version of the Scout

Fig. 1.1 The Scout D Launcher

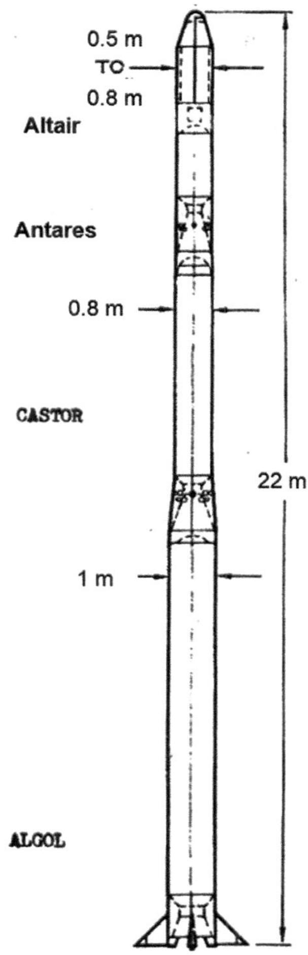

were developed during time by several upgrading of the stages. The last configuration was Scout G which consisted of the following stages: Algol3A, Castor2, Antares3A and Altair3A. This configuration will be considered in the following together with the Scout D configuration which has the same stages except that the third stage is Antares2A instead of Antares3A [1–4].

Table 1.1 gives the main flight events, the corresponding time and mass of a Scout G with no payload mass and no coasting between the stages [3]. If the spacecraft to be injected into orbit has mass m_{pay} the payload mass to be added is $m_U = m_{pay} + m_{adapter}$ (see later for the payload adapter mass).

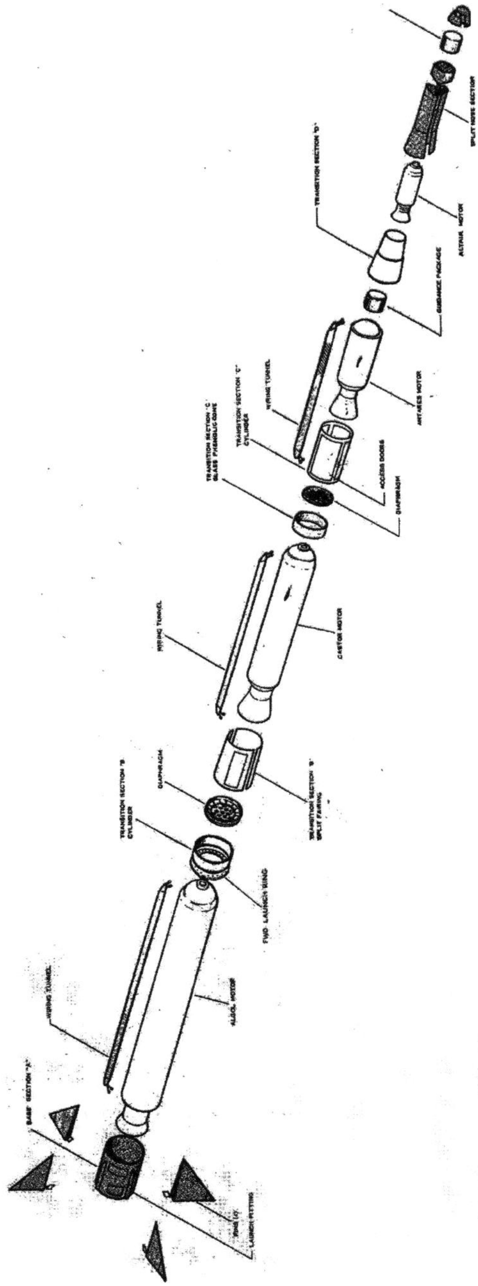

Fig. 1.2 The Scout general structure [5]

Table 1.1 Typical sequence of events during a flight: time and launcher mass variation: Scout G, no payload mass, no coasting between stages [3]

Flight events	time (s)	Vehicle mass (kg)
Ignition Stage 1	0	21189
Start timer	0.03	21184
Stage 1 burn out	80	8272
Stage 2 ignition	80.5	6242
Stage 2 burn out	119.5	2513
Separation of heat shield	123	2393
Stage 3 ignition	124.5	1698
Stage 3 burnout	172	412
Activation coast control	177	412
Spin motor ignition	612.5	412
Stage 3 separation	614	314
Retro force command	615	314
Ignition of Stage 4	619	314
Stage 4 burn out	653	28

Figure 1.3 shows the flight loads (axial acceleration computed in g) on the Scout D launcher during a nominal mission [3]. Higher loads occur for last stages, having higher thrust over weight ratio

$$n = \frac{T}{g\, m} \tag{1.1}$$

Note that zero load corresponds to the coasting (no thrust) arcs.

Figure 1.4 shows the velocity, the altitude and the range of a mission by Scout D launcher targeting a circular orbit of altitude 650 km [3]. It is apparent that the slope of the velocity curve increases with time, since n is increasing due to the rapid consumption of the propellant mass. Moreover small decrement in velocity are observed at the tail of the thrust and at stage separation. The coasting after the third stage burn out implies a a decrease in velocity as the launcher gains altitude. This is similar to what happens after the first burn of a two burns Hohmann transfer [6]. The continuous fourth stage thrust is realized instead of the second Hohmann burn to gain the required velocity, while the altitude is close to the required one. The strategy of burns and the duration of the coasting arc is aimed at maximising the payload mass that the launcher can inject into orbits of different radius and inclination. Figure 1.5 shows the payload mass that Scout can inject into elliptic orbits launching toward East from the San Marco Platform (latitude –2 deg) [3]. The lower curve shows the payload mass that can be inject on circular orbits of different radius and inclination 2 deg. Payload masses for eccentric orbits are also reported on the curves having the same pericenter and different apocenter.

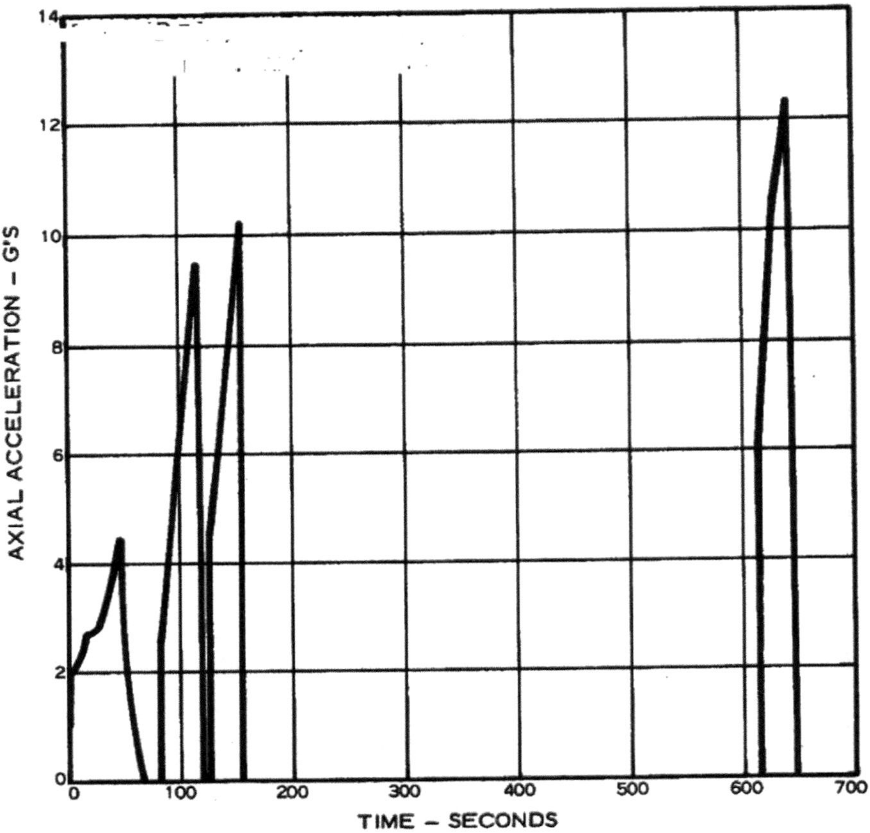

Fig. 1.3 Typical axial loads in g, [5]

1.2 Optimal Staging

The velocity of a spacecraft in a Low Earth Orbit (LEO) exceeds 7500 m/s. To achieve such orbital velocity, the launcher engines must generate an acceleration able to increase the spacecraft velocity by that amount, while compensating for the gravity and the drag effects that are opposite to the velocity gain. These adverse effects, together with others, are called losses, and will be quantified later: here only the engine contribution to the variation of velocity is computed.

Consider a rocket having a constant thrust T approximated by:

$$T = g\,Isp\;m_d$$

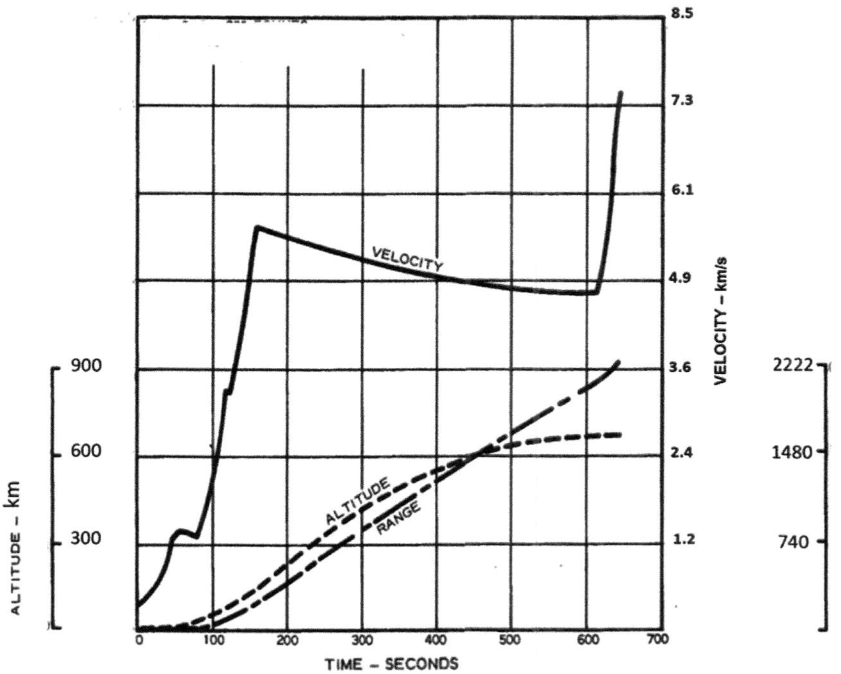

Fig. 1.4 The Scout D velocity, altitude and range, targeting a circular orbit [5]

where $(g\,Isp)$ is the velocity of the exhaust gasses leaving the nozzle, with $g = 9.81\,\frac{m}{s^2}$, Isp is called specific impulse (in seconds), and m_d is the rate of variation of the mass

$$\dot{m} = -m_d$$

due to consumption of the propellant, having total mass m_p. The acceleration that the engine can provide to the rocket is $a = \frac{T}{m}$, hence the velocity generated during the thrust time t_b is the integral of the acceleration:

$$\Delta V = V_f - V_0 = \int_0^{t_b} \frac{T}{m}\,dt \tag{1.2}$$

that is:

$$\Delta V = g\,Isp \int_0^{t_b} \frac{m_d}{m}\,dt = -g\,Isp \int_0^{t_b} \frac{\dot{m}}{m}\,dt = -g\,Isp \int_0^{t_b} \frac{dm}{m}$$

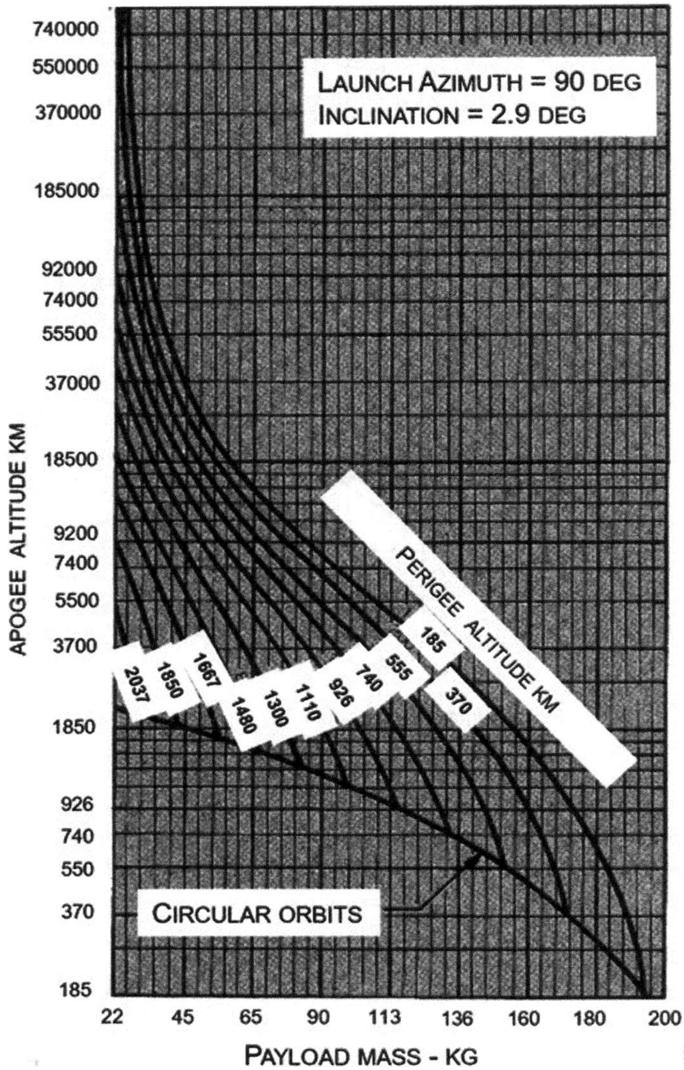

Fig. 1.5 The Scout D payload mass in elliptic orbits from San Marco launch base [5]

The result is called **Tziolkowski law**:

$$\Delta V = V_f - V_0 = g\, Isp\, log\left(\frac{m_0}{m_f}\right) \tag{1.3}$$

where m_0 is the rocket initial mass and $m_f = m(t_b)$ is the rocket mass after the propellant consumption. This is regarded as the structural mass m_s after the burn:

$$m_s = m_0 - m_p$$

The structural ratio is defined as

$$u = \frac{m_s}{m_0} \tag{1.4}$$

and the Tziolkowski law (1.3) is equal to:

$$\Delta V = g\, Isp\, log\left(\frac{1}{u}\right) \tag{1.5}$$

In (1.5) the specific impulse depends on the type of propellant selected, and u depends on the industrial technology level of the rocket system.

It has to be remarked that three parameters $\{u, Isp, t_b\}$, where t_b is the duration of the engine burn (burn out time), determine the main engine characteristics, in particular: (a) the initial thrust to weight ratio n_0, (b) the average thrust to weight ratio $\bar{n}$, c) the Tziolkowski variation of velocity ΔV_{Tz}.

In fact, during the engine burn the mass variation is well approximated by a linear function in time:

$$m = m_0 - m_d\, t$$

Then at burn out time t_b:

$$m_s = m_0 - m_d\, t_b$$

Dividing by m_0 the above relation gives:

$$u = \frac{m_s}{m_0} = 1 - \frac{m_d}{m_0}\, t_b = 1 - q\, t_b$$

with

$$q = \frac{m_d}{m_0} = \frac{1-u}{t_b}$$

The initial thrust over weight ratio is defined as

$$n_0 = \frac{T}{g\, m_0} = Isp\, \frac{m_d}{m_0} = Isp\, q = Isp\, \frac{1-u}{t_b} \tag{1.6}$$

The Tziolkowski variation of velocity is:

$$\Delta V_{Tz} = g\, Isp\, log\left(\frac{1}{u}\right) = g\, \frac{n_0\, t_b}{1-u}\, log\left(\frac{1}{u}\right) = g\, \bar{n}\, t_b \tag{1.7}$$

where

$$\bar{n} = n_0\, \frac{log(\frac{1}{u})}{1-u} \tag{1.8}$$

The constant $\bar{n}$ is in fact the average in time of the thrust over weight ratio (1.1). Namely:

$$\bar{n} = \frac{1}{t_b}\int_0^{tb} \frac{T}{g\,m}(t)\,dt = \frac{T}{g\,m_0}\frac{1}{t_b}\int_0^{tb}\frac{dt}{1-qt} =$$
$$= -\frac{n_0}{t_b}\frac{1}{q}\ln(1-qt)|_0^{tb} = -\frac{n_0}{q\,t_b}\ln(1-qt_b) =$$
$$= -\frac{n_0}{1-u}\ln(u) = \frac{n_0}{1-u}\ln\left(\frac{1}{u}\right)$$

The current space engines and materials do not allow to generate a ΔV capable to reach the orbital velocity. It is necessary to reduce the structural ratio splitting the rocket into different parts (stages) and releasing the tank of the exhaust engine before firing the next one. In this way, the ΔV provided to the spacecraft will be the sum of different contributions having better values of the structural ratio.

The optimal distribution of the mass of a launcher into several stages is now determined [7].

Consider a launcher with a number N of stages: total mass of each stage is denoted by m_k and the propellant mass is m_{p_k} , $k = 1 \ldots N$. Then a structural mass ratio is defined for each stage

$$u_k = \frac{m_{s_k}}{m_k} = \frac{m_k - m_{p_k}}{m_k} \quad k = 1 \ldots N$$

The total mass of the launcher is given by the sum of the stages plus the payload mass:

$$M_1 = \Sigma_{j=1}^{N} m_j + m_{pay} \tag{1.9}$$

After the burn out of each stage, the stage mass is separated from the launcher and a subrocket is defined with mass:

$$M_k = \Sigma_{j=k}^{N} m_j + m_{pay} \quad ,k = 2 \ldots N \tag{1.10}$$

The structural mass of each subrocket is defined by

$$U_k = \frac{M_k - m_{p_k}}{M_k} \quad k = 1, \ldots N \tag{1.11}$$

By definition

$$m_{s_k} = \frac{M_k(1-U_k)}{1-u_k} \;,\quad m_{p_k} = M_k(1-U_k) \tag{1.12}$$

$$M_k = M_{k-1}\frac{(U_{k-1} - u_{k-1})}{1-u_{k-1}} \;,k = 2, \ldots, N-1 \;,\quad M_N = m_{pay} \tag{1.13}$$

Let the values M_1, m_{pay} be given, then

$$\frac{m_{pay}}{M_1} = \Pi_{i=2}^{N} \frac{M_i}{M_{i-1}} \tag{1.14}$$

The Tziolkowski velocity is equal to

$$\Delta V_{Tz} = -g \Sigma_{i=1}^{N} Isp_i \, \log(U_i) \tag{1.15}$$

The problem is now: select the values $\{U_k^*\}$ to maximize the Tziolkowsky ΔV, corresponding to minimize the opposite of (1.15)

$$J = g \Sigma_{i=1}^{N} Isp_i \, \log(U_i)$$

under the constraint (1.14) which is equal to (see (1.13)):

$$\Phi = \frac{m_{pay}}{M_1} - \Pi_{i=2}^{N} \frac{U_i - u_i}{1 - u_i} = 0 \tag{1.16}$$

To solve the constrained minimum problem, the Hamilton function is introduced with lagrangian multiplier λ:

$$H = J + \lambda \, \Phi$$

and the solution for the N+1 unknowns $\{U_i, \lambda\}$ derives from the N+1 equations

$$\begin{aligned} \frac{\partial H}{\partial U_i} &= 0 \\ \frac{\partial H}{\partial \lambda} &= 0 \end{aligned} \tag{1.17}$$

These equations correspond to

$$\frac{g Isp_i}{U_i} - \frac{\lambda}{1 - u_i} \Pi_{k=1, k \neq i}^{N} \frac{U_k - u_k}{1 - u_k} = 0 \tag{1.18}$$

and to Eq. (1.16). From (1.18) the lagrangian multiplier λ is derived:

$$\lambda = \frac{g Isp_i (U_i - u_i)}{U_i} \Pi_{k=1}^{N} \frac{1 - u_k}{U_k - u_k} \tag{1.19}$$

Equation (1.19) holds true for any $i = 1, N$, this corresponds to

$$\frac{g Isp_1 (U_1 - u_1)}{U_1} = \ldots \cdots = \frac{g Isp_N (U_N - u_N)}{U_N}$$

that is

$$\frac{gIsp_i(U_i - u_i)}{U_i} = \frac{gIsp_1(U_1 - u_1)}{U_1}, i = 2, N \tag{1.20}$$

Equation (1.20) corresponds to

$$U_i = \frac{Isp_i\, u_i}{U_1(Isp_i - Isp_1) + Isp_1\, u_1} U_1, i = 2, N \tag{1.21}$$

Replacing the formulas (1.21) into (1.16) one has the following algebric equation of order N in the unique unknown U_1:

$$\frac{m_{pay}}{M_1} \Pi_{i=1}^{N} \left[U_1(Isp_i - Isp_1) + Isp_1\, u_1\right] - \Pi_{i_1}^{N} \frac{u_i}{1 - u_i} \left(Isp_1(U_1 - u_1)\right)^N = 0 \tag{1.22}$$

The solution U_1^* of (1.22) is used in (1.21) to obtain all the subrocket structural mass ratios $\left\{U_i^*\right\}_{i=1,N}$ maximizing the Tziolkowski variation of velocity ΔV_{Tz}.

1.3 Staging of Scout and of Vega Launchers

This section compares the optimal staging with respect to the actual staging of the Scout and Vega launchers.

The Scout launcher has four stages which are described hereafter.

1.3.1 Scout First Stage

The Scout first stage is called Algol 3A and it consists of a case, containing the propellant to be burned inside the combustion chamber, the nozzle and an additional section, called Section A, located at the base of the launcher supporting the external wings and jet vain for the launcher control during the first stage burn. The stage total mass is equal to 14215 kg, the propellant mass is equal to 12712 kg [10]. In addition to the propellant, some additional mass is consumed during the combustion, in particular the liner used for thermal protection of the tank and some elements of the nozzle: this extra mass consumed is equal to 98 kg. The stage dimensions (in mm) are shown in Fig. 1.6. The stage total length is 9.61 m, the outer diameter is equal to 1.142 m and the inner diameter (combustion chamber) is on average equal to 1.133 m. When fully loaded, the center of mass of the first stage is located on the axial axis at 4.43 m from the top of the stage. At the end of the combustion the center of mass is distant 6.18 m from the top.

The case is a steel cylinder of length 7.95 m, outer diameter of 1.142 m and an average structure width equal to 1 cm. The forward and aft closure are of semi-elliptic shape: an igniter boss is located on the forward closure and the aft end is machined to host the nozzle.

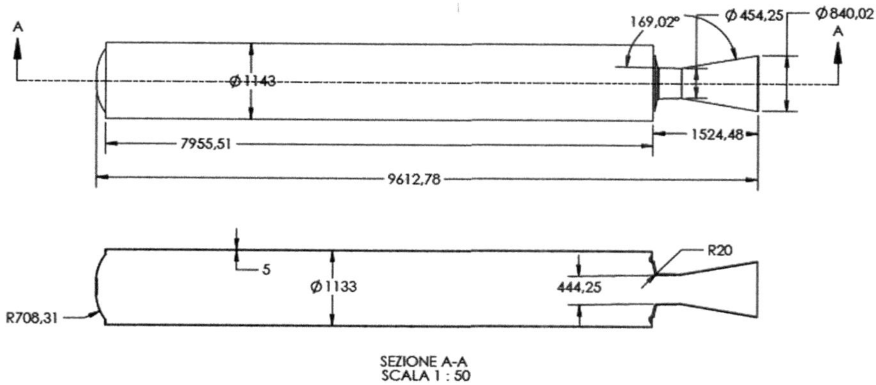

Fig. 1.6 Dimensions (in mm) of the first stage Algol, [5]

The nozzle is made of steel and it has a machined graphite insert in the throat. The nozzle lenght is equal to 1.52 m. The nozzle area at the exhaust section is 0.53 m^2 (radius at exit 0.411 m) and the area of the nozzle's throat is equal to 0.087 m^2 (throat radius 0.161 m), A shear ring and a two place retainer ring secure the nozzle to the aft end of the case.

1.3.2 Scout Second Stage

The second stage is called Castor 2. The stage total mass is equal to 4433 kg with propellant mass equal to 3724 kg. The stage dimensions (in mm) are shown in Fig. 1.7. The stage total length is 6.19 m; when it is full loaded the center of mass of the first stage is located on the axial axis at 2.57 m from the top of the stage.

The case is a steel cylinder of length 5.10 m, 0.787 m of external diameter and width equal to 1 cm on average. Circular bolt-flanges are welded to the forward and aft sections (of semi-elliptic shape): they are used to attach the motor to the veicle frame. An igniter boss is located on the forward closure.The inside of the case is lined with a rubber—like compound and then filled with solid propellant [10].

The nozzle is fabricated of steel and it has a convergent-divergent cone type geometry. Its weight is about 227 kg. A graphite insert is pressed into the throat area and an insulated material is located between the graphite and the nozzle itself. The nozzle length is equal to 1.08 m. The nozzle area at the exhaust section is 0.81 m^2 and the area of the nozzle's throat is equal to 0.039 m^2.

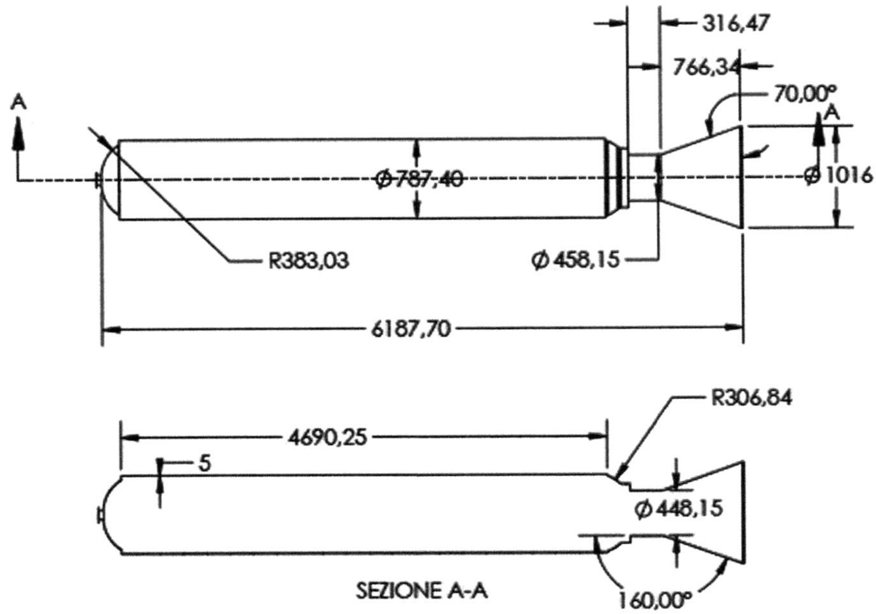

Fig. 1.7 Dimensions (in mm) of the second stage Castor, [5]

1.3.3 Scout Third Stage

The third stage is called Antares 2A: its total mass is equal to 1269 kg with propellant mass equal to 1161 kg. The stage dimensions (in mm) are shown in Fig. 1.8. The stage total length is 2.89 m.

The case has length 2 m, external diameter of 0.76 m and average structure width of 2.5 cm. It is made of filament winding glass fiber reinforced epoxy resin structure and incorporates integrally wound high-strength aluminium forward and aft adapter. The forward and aft ends are semi-elliptical in shape: the forward end has an igniter boss and the aft end is machined to receive the nozzle.

The nozzle has length of 0.96 m and it is made by filament winding with two layers of glass fibre. The nozzle area at the exhaust section is 0.43 m^2 and the area of the nozzle's throat is equal to 0.025 m^2 [10].

1.3.4 Scout Fourth Stage

The fourth stage is called Atair 3A, the total mass is equal to 301 kg with propellant mass equal to 273 kg. The stage dimensions (in mm) are shown in Fig. 1.9. The stage total length is 1.48 m, the outer diameter is equal to 0.5 m.

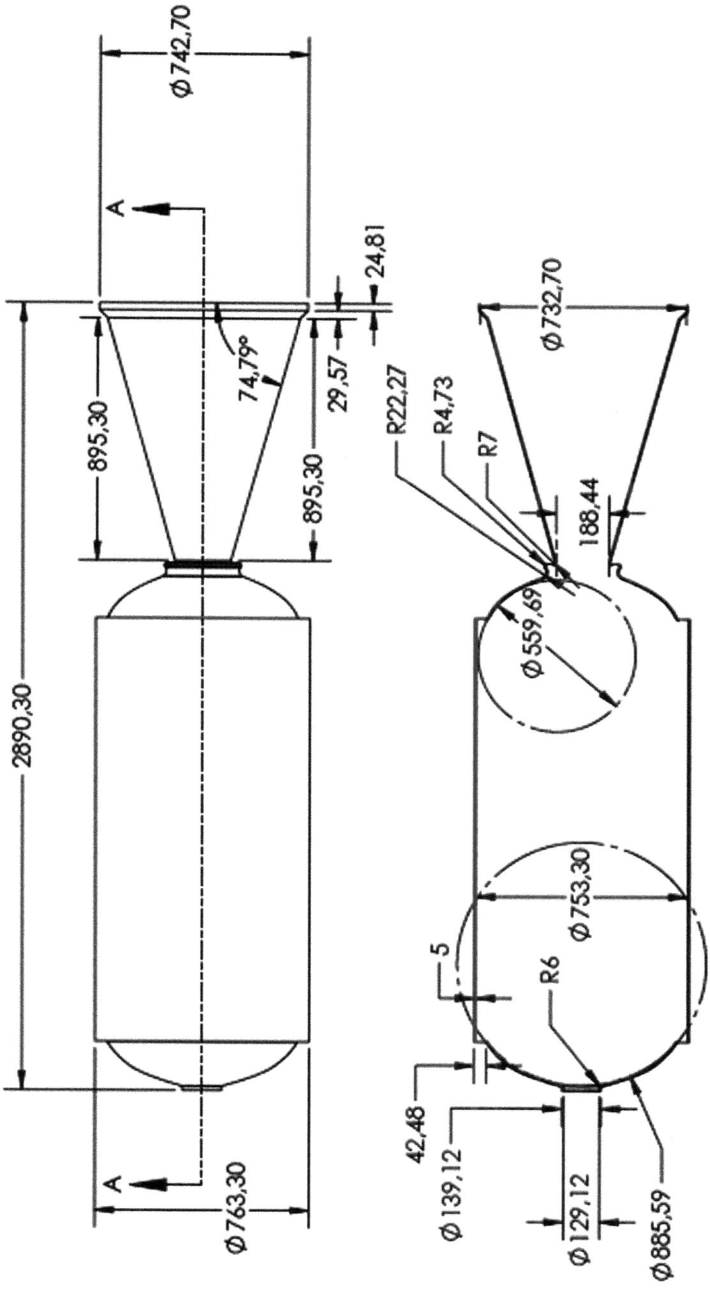

Fig. 1.8 Dimensions (in mm) of the third stage Antares 2A, [5]

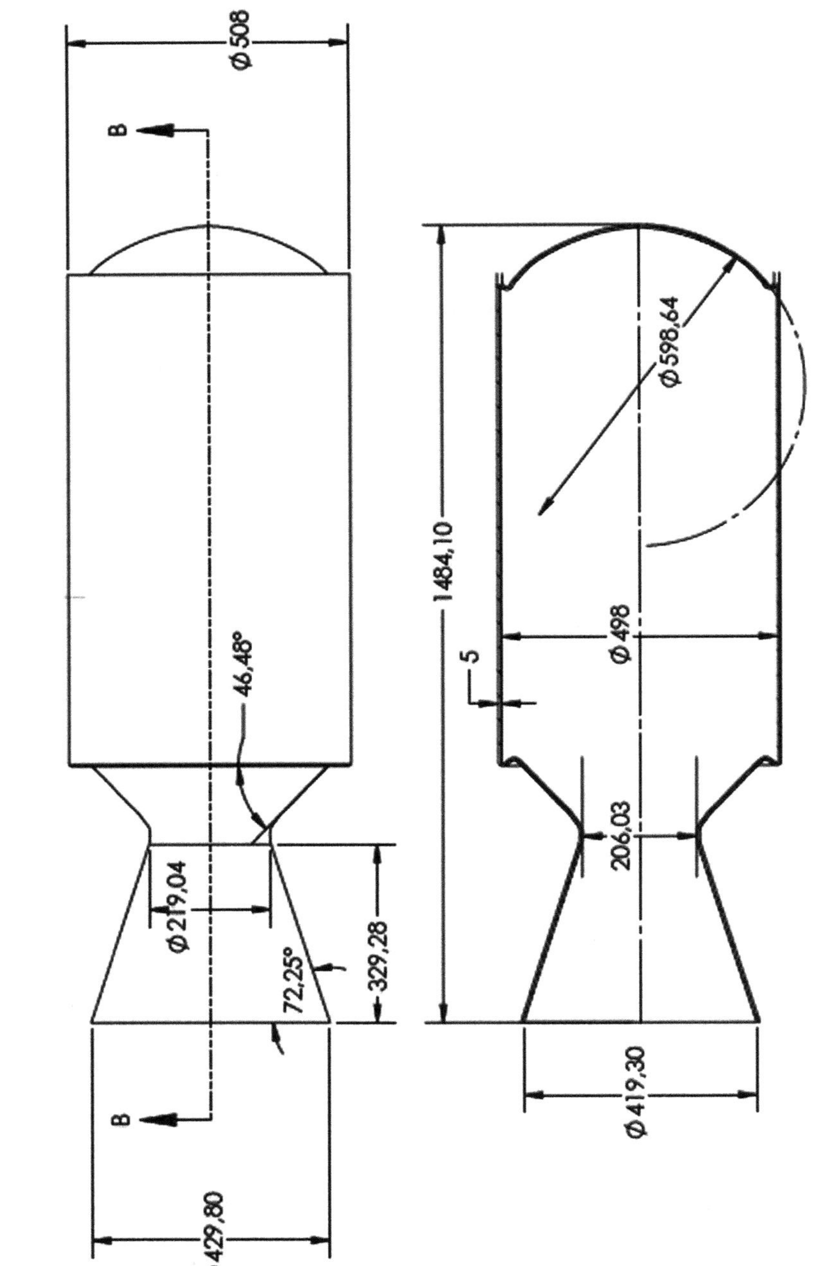

Fig. 1.9 Dimensions (in mm) of the fourth stage Castor, [5]

The case is a 1.15 m cylinder fabricated of helically wound epoxy resin impregnated fibreglass. Aluminium forward and aft adapters are integrally wound into the hemispherical ends of the case. The chamber is wound directly over an insulator of silica filled Buna-N rubber.

The nozzle consists of an asbestos phenolic molded retainer, a carbon throat insert, and an exit cone with a torque ring on the aft end. The nozzle length is equal to 0.33 m. The nozzle area at the exhaust section is 0.139 m^2 and the area of the nozzle's throat is equal to 0.003 m^2. The nozzle is attached to the motor aft adapter by 15 retaining bolts [10].

1.3.5 Payload Adapter and Heat Shield

Different payload adapter are used depending on the payload mass. Table 1.2 defines the different adapters and their mass. Different heat shields are used for different payloads. They all consists of fibreglass honeycomb sandwich with a stainless steel nose cap. The heat shield is released and jettisoned by the use of ballistic actuators located along the longitudinal axis. Namely the heat shield is fabricated by two half shells. The nose cap is attached to one of the two shells and butts against the other half shell. The separation plane lines along the pitch axis and the separation line has shear clips and dovetails attached to each half shell to accomplish shear and momentum continuity across the separation plane. The two half shells are joined by a ballistic actuator at the top, a band on the cylindrical section and a Marmor clamp at the base. A bumper between the heat shield and the motor prevents the heat shield from hitting the payload under high load conditions and transfers lateral inertial loads from the motor case to the heat shield. The heat shields are ejected at the third stage ignition signal. In fact the band at the cylindric section is released at the second stage burn out by a command to fire the explosive bolts on the band. Ejection is achieved by simultaneous firing of the explosive bolts on the Marmor clamp and the ballistic actuator [11].

Table 1.2 Payload adapters

Payload mass (kg)	Adapter name	Adapter mass (kg)
100	EG	6
220	E	8
290	200E	9
>290	25E	10

1.3.6 Scout Optimal Staging

Table 1.3 summarizes the mass data of each stage of the Scout G configuration [4] and Table 1.4 provides the engine data. Table 1.5 gives the subrocket data considering: (i) 150 kg of payload, (ii) 8 kg for the adapter, (iii) 20 kg of avionics, (iv) 120 kg for the heat shield. Each subrocket mass is computed by formulas (1.9), (1.10), the structural ratio U_k is computed according to (1.11) the initial thrust over weight ratio n_0 is computed according to (1.6), the average thrust over weight ratio $\bar{n}$ are computed according to (1.8) Tziolkowski velocities provided by each stage are computed according formula (1.15).

The total the variation of velocity that the G Scout launcher engines can generate is then: $\Delta V_{tot} = 10213\,\text{m/s}$.

This result is compared to the optimal mass distribution that can be obtained by the formula of Sect. 1.2 (same payload mass, adapter and heat shields). This gives the results of Table 1.6 showing the optimal mass of each stage and the Tziolkowski ΔV. The total ΔV produced by the optimal staging is 10521 m/s, with an increase of about 300 m/s over the real configuration. In particular the third stage performance is improved at expense of the other stages. Note that the Scout launcher was originally designed by assembling existing engines and that the main improvements during its 40 years of service were mainly devoted to the updating of the third stage.

Table 1.3 Scout G mass of each stage

Stage	Structure mass m_s (kg)	Propellant mass m_p (kg)	Total mass m (kg)	ratio $u = m_s/m$
Stage 1	2030	12817	14847	0.137
Stage 2	695	3729	4334	0.157
Stage 3	98	1286	1384	0.071
Stage 4	28	286	314	0.089

Table 1.4 Scout G engine average main values

Stage	Average thrust (KN)	Specific impulse (s)	Burn time (s)
1	404.18	257	80.0
2	264.4	280	38.8
3	77.7	293	47.6
4	18.0	218	34.0

Table 1.5 Scout G subrockets with payload and adapter 150 kg, heat shield 120 kg

Launcher	M_k kg	M_{k_f} kg	U_k	n_0	$\bar{n}$	ΔV_{Tz}m/s
SubRocket 1	21239	8422	0.396	1.94	2.97	2333
SubRocket 2	6392	2663	0.416	4.21	6.33	2409
SubRocket 3	1848	562	0.304	4.28	7.33	3423
SubRocket 4	464	178	0.384	3.95	6.14	2048

Table 1.6 Scout G optimal distribution of mass and optimal Tziolkowki velocities

Stage	M_k	Propellant mass m_p (kg)	Structure mass m_s (kg)	ΔV m/s
Stage 1	21239	11531	1826	1974
Stage 2	7882	4410	882	2252
Stage 3	2649	2162	165	4869
Stage 4	322	157	15	1426

Table 1.7 Data of the Vega stages

Stage	m_{tot} (kg)	m_p (kg)	Isp (s)	u_k	$\bar{T}$ (KN)
Stage 1	96246	87710	280	0.094	3015
Stage 2	26300	23814	287	0.113	1122
Stage 3	12000	10567	296	0.119	314
Stage 4	955	572	314	0.401	2.42

1.3.7 Vega Launcher

Another example of a space launcher is presented: Vega is the last launcher of the European Space Agency, produced under the leadership of the Italian company Avio to complement the market of the Ariane Space Group with a small launcher. The reference orbit is a polar orbit from Kourou space base and payload of 1500 kg [8]. The basic data of the Vega stages can be found in the Table 1.7.

With a payload of 1500 kg, adapter of 77 kg and heat shields of 540 kg the data of Table 1.8 are obtained.

Table 1.8 Vega data with payload of 1500 kg, adapter of 77 kg and heat shield of 540 kg

Stage	$M_k(kg)$	M_{kf} (kg)	n_0	$\bar{n}$	U_k	ΔV_{Tz}(m/s)
Stage 1	137615	49908	0.363	1.62	2.58	2786
Stage 2	41372	17558	0.424	2.14	3.19	2413
Stage 3	15072	4505	0.299	1.77	3.05	3771
Stage 4	2532	1960	0.774	0.10	0.11	789

Table 1.9 Optimal staging for Vega

Stage	M_k (kg)	Propellant mass m_p (kg)	ΔV
Stage 1	137615	98700	3470
Stage 2	28710	19379	3164
Stage 3	6863	4633	3264
Stage 4	1602	15	29

The total (Tziolkowski) variation of velocity of the Vega launcher is then $\Delta V_{tot} = 9495\,\mathrm{m/s}$.

The optimal staging for Vega and the related Tziolkowsi velocities are given in Table 1.9. The optimal variation of velocity is $\Delta V_{total} = 9927\,\mathrm{m/s}$. Note that the mass of the fourth stage is reduced and the optimal staging basically suggests to consider a three stage rocket, as it is the evolution of the Vega launcher, Vega E [9].

1.4 Separation System

Staging is actuated by the launcher separation system. The separation command is given by the on board computer according to a logic that takes into account the time elapsed from the lift off and the ignition of the successive engines and the data from the accelerometers measuring the thrust acceleration. When the expected burn time has elapsed and the thrust is below a threshold value, the rocket guidance program is stopped and the attitude is kept constant so that the angular velocity is zero. The separation command is then given.

Separation between two stages can take place in two ways: cold or hot separation. In the cold separation the command activates a small pyro charge that disengages a mechanical device that keeps the two stages coupled. A typical device is the so called Marmor clamp. This is a strap tensioned by pyro fasteners around the interface of two flanges to hold them together.

In the hot separation system the ignition of the upper stage removes the mechanical device that holds the two stages together.

The Scout launcher has both hot and cold separation systems. The first stage separation system is located in the interstage connecting the first to the second stage, called B section. This section is divided into two parts, called lower B and upper B section. These two sections are connected by a frangible diaphragm, see Fig. 1.10. Namely the diaphragm is screwed to both the sections which contain the threads located on the adaptor rings. Upon second stage ignition, the diaprhagm is ruptured, then the diaphragm threads disengage the second stage [11]. Figure 1.11 describes the stage separation.

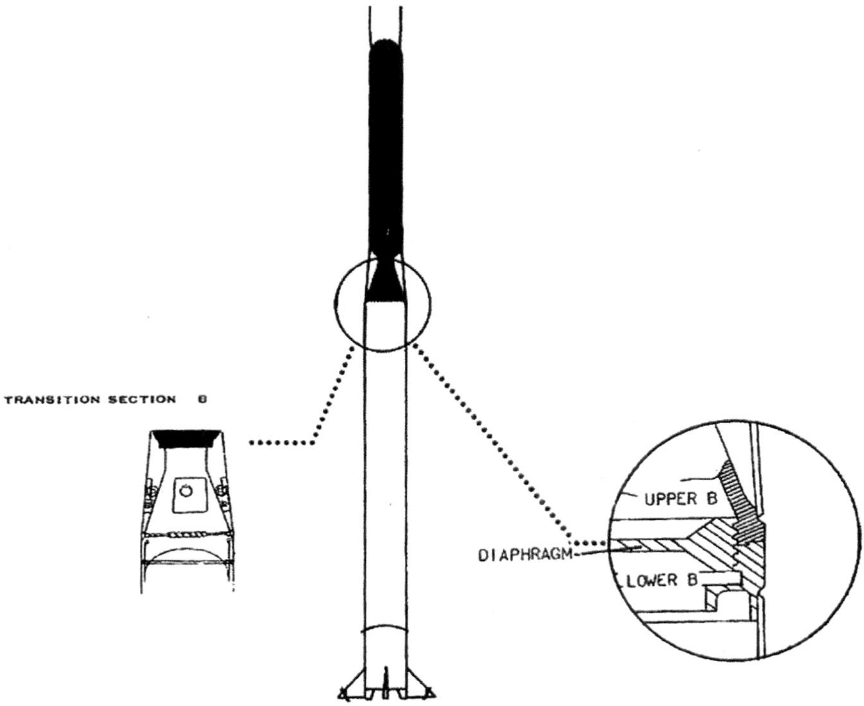

Fig. 1.10 The diaphragm connecting the first and second stages [5]

The second stage separation system is located in the interstage between the second and third stages, called Section C. This is also divided into Lower and Upper C. The general assembly is the same of the Interstage 1–2: Lower C is dedicated to the separation system, Upper C contains the guidance and control system of the third stage. The second stage separation system is again provided by a frangible diaphragm [11]. Figure 1.12 describes how the diaphragm connects the Lower and Upper C sections.

Finally the third stage separation system is located in the last interstage. This is composed by Lower and Upper Section D. Lower section D is attached to the top of the third stage and supports the 4 stage spin bearing. It also contains the guidance equipment. Upper D is attached to the four stage motor; its rear end is attached to the bearing ring which contains the spin bearing and the separation diaphragm. Upper D contains 4 spin motors which are actuated just before the separation between the third and fourth stages. Approximately 4.5 s before the ignition of the four stage motor, a four piece Marmor clamp, held together by four explosive bolts, is used for the separation. Mechanical springs mounted below the separation point push the fourth stage away from the rest of the vehicle. As the springs push upwards, the vehicle timer sends a signal to operate all the four third stage reaction control

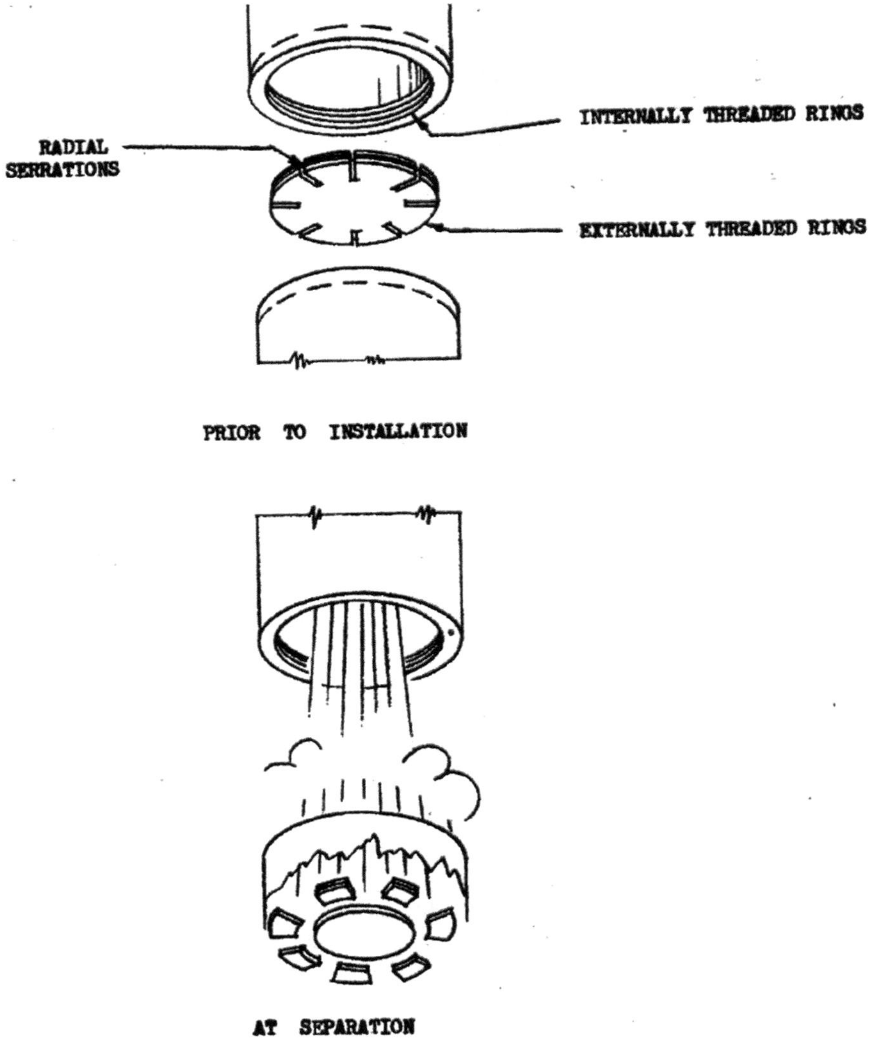

Fig. 1.11 Stage separation system [5]

motors simultaneously. As these motors are slightly tilted upwards, their simultaneous operation will augment the separation by pushing the third stage downwards. The fourth stage is then separated from the rest of the vehicle before its ignition [11]. Figure 1.13 shows a detail of the separation system.

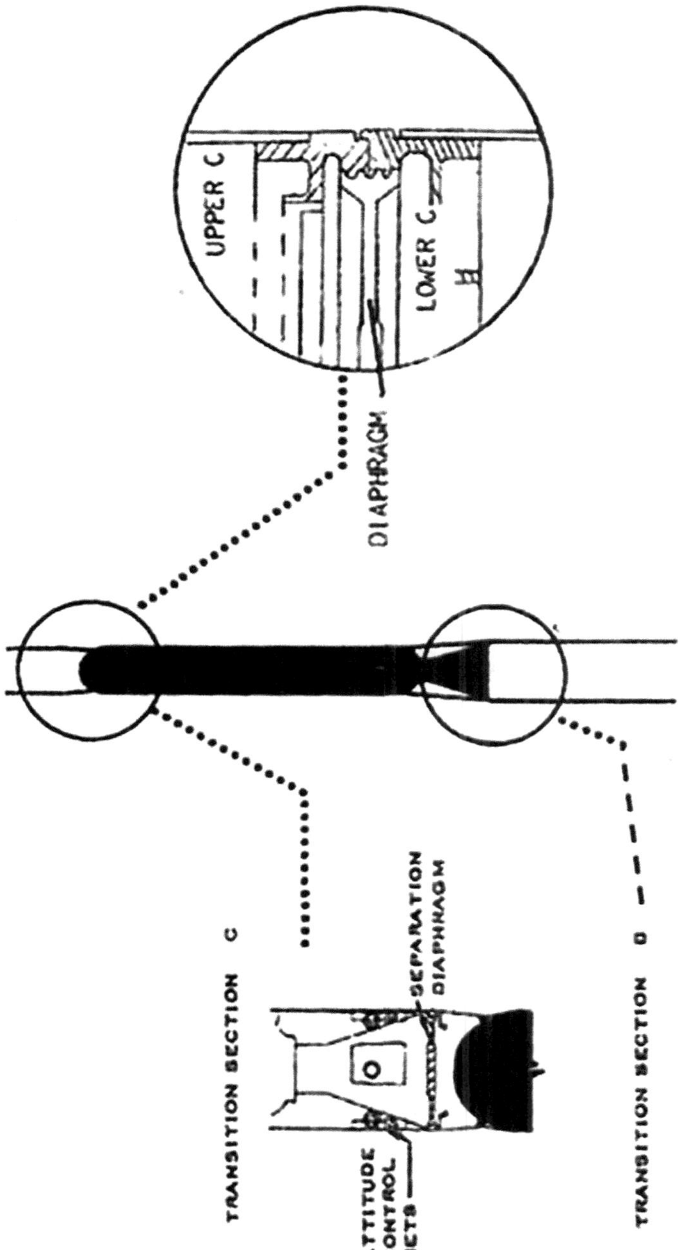

Fig. 1.12 The separation system of the second stage [5]

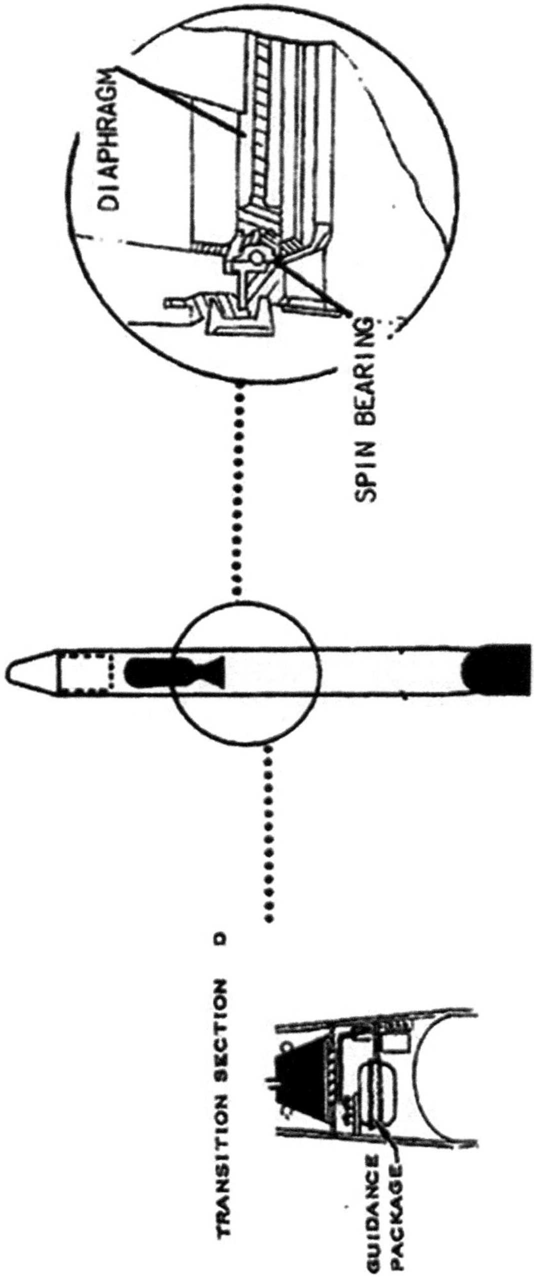

Fig. 1.13 The separation system in the section D [5]

1.5 Propulsion System

Rocket engines can be divided into three categories: liquid monopropellant or bipropellant, solid propellant and hybrid engines.

Liquid monopropellants use a fuel that spontaneously decomposes in the combustion chamber under certain conditions.

Liquid bipropellant rockets mix two reactants (oxidiser and fuel) required for combustion. These are stored separately in gas or liquid form and then mixed when thrust is required.

With solid propellants the two reactants are mixed in a solid grain.

Hybrid engines use a solid fuel grain with a single plumbing system to inject the oxidizer when the thrust is required.

The best performance in terms of thrust is achieved with liquid propellants, but they need a control network between the oxidiser and the fuel which requires plumbing components, adds extra weight and increases the complexity compared to solid propellants. In addition, liquid propellants are less stable than solid propellants in critical situations, such as before and during the take-off. On the other hand with liquid propellants the thrust can be controlled and can be shut down and restarted as needed, whereas solid propellant combustion cannot be stopped.

This leads to the design of hybrid engines: the thrust can be initiated, regulated and stopped. The technological development of space launcher with hybrid engines is still in progress.

The basic parameters for solid propellant engines are:

- Specific Impulse (in seconds) $Isp = \frac{T}{g\, m_d}$, with T thrust force and m_d is the rate of propellant consumption
- Burning rate r_b (in millimeters over seconds)
- Density ρ_p
- Grain geometry
- Propellant mass m_p
- Burn out time t_b.

A first dimensioning of the rocket engine is given: the case diameter will be the sum of structure thickness, propellant charge diameter $d_p = r_t\, t_b$ plus the dimension of the cross section area of the central core depending on the grain geometry. The length of the propellant charge is equal to $l = 4\frac{Vol}{\pi d^2}$, with $Volume = \frac{m_p}{\rho_P}$.

The Scout launcher has three different solid propellants for its four engines [10]. The Scout first stage engine, called Algol, has a cast-in-case aluminized composite propellant with a polybutadiene acrylic nitrile (PBAN) binder. The main characteristics of the propellant material are summarised in the following Table 1.10. The grain geometry defines the burning surface of the propellant. If the bore channel inside the propellant charge is cylindrical, the burning surface will increase as the propellant is consumed, increasing the diameter of the bore at the rate r_b. The increase in burning surface produces more thrust, so the thrust will increase over time. In general differ-

Table 1.10 Propellant PBAN main characteristics

Propellant density	1755 kg/m^3
Grain regression velocity	5.89 mm/s
Specific heat ratio	1.18
Molecular weight	20.11

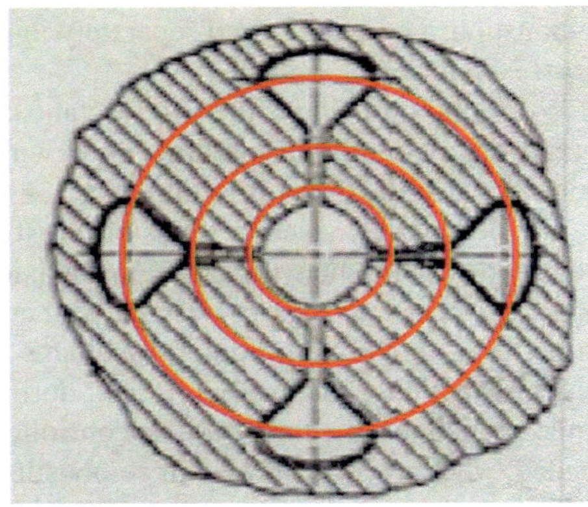

Fig. 1.14 Grain geometry of the Algol engine [5]

ent grain geometries are designed: for example a star-shaped grain keeps the burning surface, and hence the thrust, almost constant.

The grain geometry of the Algol engine is a tapered core four point star internal burning configuration of Fig. 1.14. The flame temperature of the burning propellant is 2650 C and the auto-ignition temperature is 340 C for the propellant and 230 C for the igniter [10].

The engine of the second stage, called Castor, has solid propellant composed by carboxyl-terminated polybutadiene (CTPB). The main properties of the propellant material are summarized in Table 1.11 [10].

The grain geometry consists of a cylinder with two radial slots.

The third stage engine, called Antares, has a cast composite double base propellant. This kind of propellant is no longer used because it is rather dangerous to handle. The main properties of the propellant material are summarised in the Table 1.12 [10]. The grain geometry is a case bonded slotted cylinder. Figure

Table 1.11 Propellant CTPB main characteristics

Propellant density	1790 kg/m^3
Grain regression velocity	8.9 mm/s
Specific heat ratio	1.16
Molecular weight	30.3

Table 1.12 Double Base Propellant main characteristics

Propellant density	1760 kg/m^3
Grain regression velocity	12.3 mm/s
Specific heat ratio	1.15
Molecular weight	19.30

1.15 shows the grain geometry: the maximum diameter (at different sections) of the circular hole is 23 cm and the width of the slots is equal to 4 cm [10].

Finally the fourth stage engine, called Altair, uses the same propellant of the first stage: (PBAN) with properties reported the Table 1.10.

The grain geometry is a cylinder with a circular perforation and one transverse slot [10].

Figure 1.16 shows the thrust of each of the four engines (Scout D) and the variation with respect to time.

1.6 Safe and Arm

The Safe and Arm system is divided into two distinct parts: one of them is related to the safe and controlled ignition of the rocket and the other related to the controlled destruction of the rocket for safety purposes [12].

1.6.1 Ignition System

Several requirements must be met prior to the ignition and start up of each engine of a space launcher. In particular the following measures are taken to reduce the risk:

- Range monitoring:
  The arrangements, procedures, and instrumentation system used during the countdown to ensure that the land, sea and airspace of the range are clear for launch.
- Countdown Procedures:
  Provides an orderly, step-by-step script to distribute in time the amount of real-time judgement required.
- Launch Commit Criteria/Mission Rules:
  Criteria and/or rules that must be met before final approval for launch.

A space rocket is in Safe state if mechanisms are in place that physically prevent the initiating charge from propagating to the engine combustion chamber. If the mechanism is modified in the Arm state, ignition can be reliably and reproducibly propagated to the energy release system.

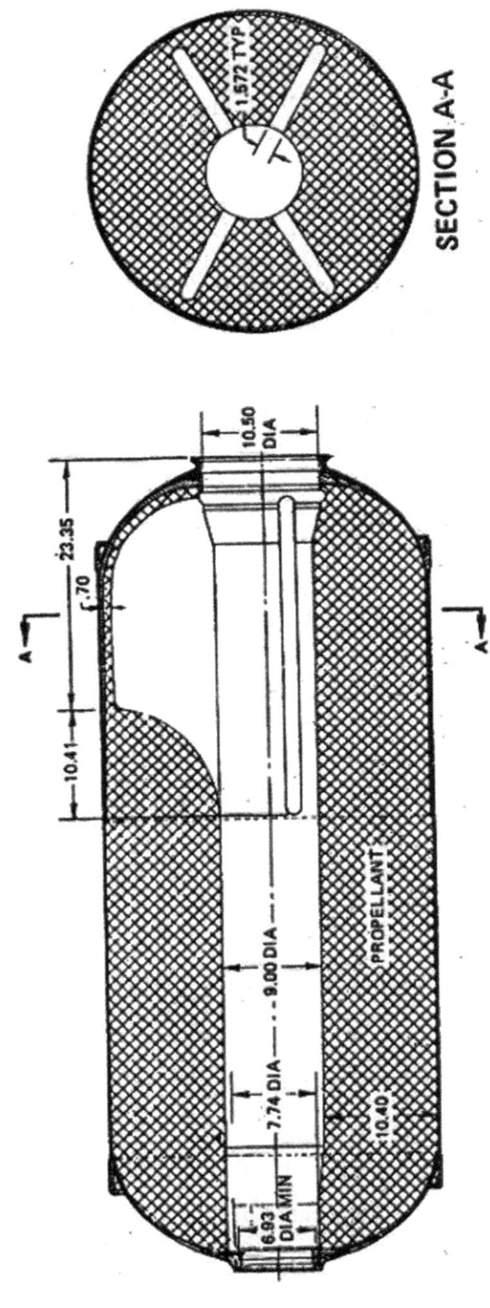

Fig. 1.15 Grain geometry of the Antares engine [5]

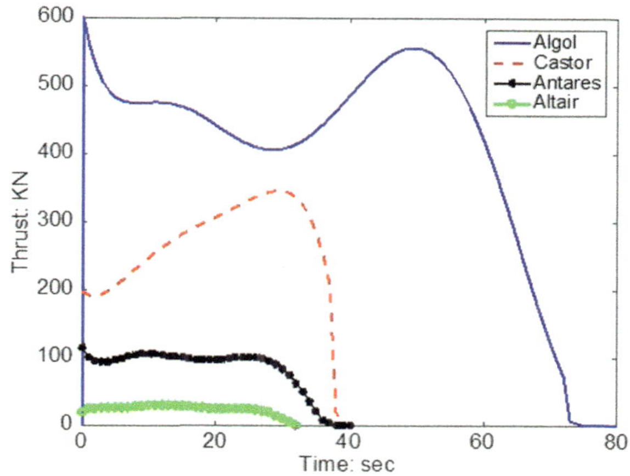

Fig. 1.16 Thrust (in KN) of the four Scout engines

The safe condition is normally achieved either by providing a barrier that can be inserted between successive elements of the ignition train or by moving elements of the train so that they are sufficiently misaligned to prevent propagation. The movement in both cases the movement can be rotary or linear, the choice usually depending on the source of actuation and the space limitations. Most mechanisms are remotely operated by electric motor or solenoid. However, mechanisms that receive their arming command from launch loads or from environmental conditions such as acceleration and altitude have been used successfully.

When the mechanical barrier is removed, the ignition system is in arming state: the electrical short circuit of the initiator is released allowing the power to be supplied by a further command to activate the ignitors.

To prevent accidental or premature arming, the ignition system shall prevent arming during and after exposure to: electromagnetic radiation, electrostatic discharge, electromagnetic pulse, electromagnetic interference, lightning effects or power supply transients. Induced currents generated by the power radiated by communications and radar equipment are of concern and one approach to protecting the ignition system is to enclose the complete equipment, including all enclosed circuits and switching and arming mechanisms, in a conductive shield. Electrostatic hazards are minimised or eliminated through the combined use of selective insulation techniques, electrically conductive bypass features, external bleed mechanisms, and non-conductive ignition materials.

System hazard analyses, such as fault tree analyses, failure mode effects and criticality analyses, are performed to estimate the failure rate of the safety system and to identify any single point, common mode, or other credible failure modes that could lead to inadvertent or premature arming or propellant firing. These analyses shall include an assessment of the relative sensitivity of each component in the pyrotechnic train. For ignition system containing an embedded computer, microprocessor, microcontroller or other computing device, the analyses shall include a determination of the contribution of the software, firmware, or microcode to the activation of a safety feature. In particular, if the software is shown to directly control safety features, a detailed analysis and testing of the applicable software must be performed to ensure that no design flaws, credible software failures, or credible hardware failures propagating through the software can result in compromise of the safety features.

Ignition of a solid rocket motor is achieved by two main ignition systems: the pyrogenic igniter and the pyrotechnic igniter. The pyrogenic igniter is very similar to a solid rocket motor but on a much smaller scale: a solid propellant is burned and provides the energy required to ignite the rocket engine. Pyrogen igniters can only work once. Pyrotechnic igniters include a wide range of different devices. They are all the devices containing a pyrotechnic composition used primarily to ignite other, more difficult to ignite materials and are electrically controlled.

Ignition systems for liquid rocket engines can be based on a variety of methods:

- Pyrogenic device:
  Due to their solid propellants, their combustion gases can generate the presence of particles in the combustion chamber: this is sometimes not allowed in liquid propellant engines.
- Spark torch
  Gaseous propellants are fed to an igniter, which ignites the propellant by an electrical spark. The combustion gases flow through a burner into the combustion chamber. Such a system requires high voltage electronics to generate the spark. A spark torch can be reignited many times making it ideal for restartable liquid propellant engines. Gaseous propellants burn cleanly and produce limited to no particles in the combustion gases. These systems however are more expensive compared to a pyrogen igniter.
- Catalytic
  An oxidant is decomposed using a catalytic bed. The decomposed gases are hot enough to ignite a fuel (which is fed to the igniter) but the combustion gases from that igniter are hot enough to ignite the engine. The fuel can be liquid or gaseous. The catalytic igniter can be simpler and cheaper than a spark torch system, however the catalytic bed will deteriorate after repeated starts.
- Hypergolic
  Two fuel combinations react spontaneously and strongly exothermically when they come into contact.

To achieve high reliability, the Scout launcher uses dual ignition systems. Safety features have been incorporated into the systems to prevent accidental or premature firing of the rocket motors. Double squibs are used in all dualised ignitors. Each one of the dual squibs is on a separate circuit and connected to a separate battery so that a failure of one electrical component will only affect one circuit.

The first stage engine is ignited by a direct electrical signal from the launch blockhouse command. Second, third, and fourth stage ignition and heat shield ejection are controlled by the guidance program timer.

Figure 1.17 shows the diagram of the Safe and Arm system of the Scout launch vehicle.

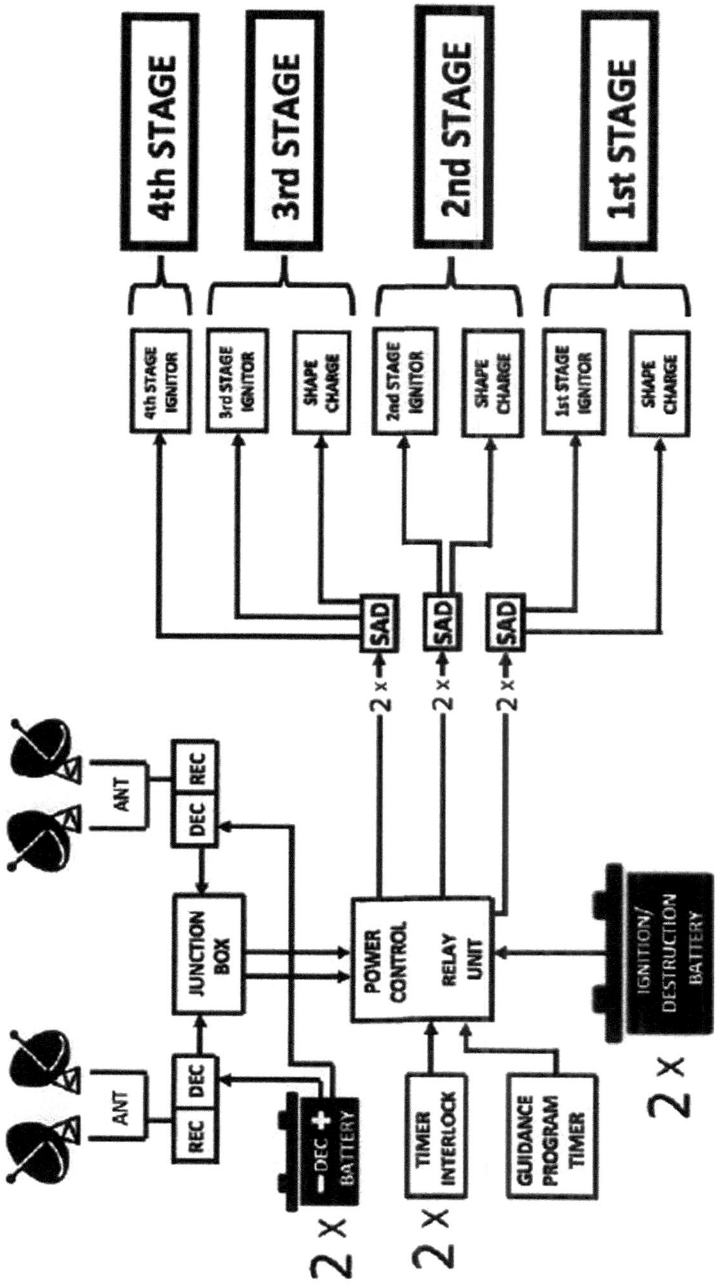

Fig. 1.17 Scheme of the Safe and Arm system, [13]

To arm the system, the flyway connection is opened by the launch blockhouse command. This feeds the Power Control Relay Box: the circuit to arm the system is still open. The circuit is closed when a "Motion Switch" activates the Ignition-Destruction Battery, located in Base A of the rocket: the launcher is armed. Then the Range Safety Officer gives the electronic ignition command by closing the switch located in the SAD (Safe and Arm Device) of the first stage and the firing command goes to the squib circuits. The SAD switches of the following stages or heat shield are activated in sequence according to the Guidance Program Timer.

In modern launchers the Flight Program Software gives the ignition commands according to the flight data processed on board.

1.6.2 Destruction System

Figure 1.17 shows that the destruction system is activated by the same S & A system used for firing: the destruction command is given from the ground by the Range Safety Officer (RSO) and received by the antennas located on the launcher.

To assist the RSO in making a decision to abort the flight, there are many indicators that show the condition of the launcher in flight. These include booster chamber pressures, vertical plane charts (later replaced by computer generated destruct lines), and altitude and velocity indicators. The RSO is supported with this information by a team reporting from profile and horizontal parallel wires used at lift-off (before radar could track the rocket) and telemetry indicators. After initial lift-off, flight information is acquired by X and C-band radars, and S-band telemetry receivers from vehicle-mounted transmitters.

The destruct command activates detonation chords running in tunnels along the longitudinal axis of the launcher, see the Launch Operations in Sect. 1.9. When they are ignited, the charges cut open the booster casing. This causes the solid propellant to burn up rapidly by exposing a large reaction surface. Moreover the cut-open casing also allows the combustion gases to escape sideways instead through the nozzle, so the booster no longer produces any significant thrust.

For liquid propellants an ARM switch disables propulsion, and the DESTRUCT signal ignites the primacord around the fuel tanks. The primary effect of the charges is to rupture the propellant tanks down the centre and spill their contents. For boosters with cryogenic propellants, the system is designed to rupture the tanks in such a way as to minimise propellant mixing, which would result in an extremely violent explosion, specifically by causing the charges to split the sides of the tanks open like a zipper, spilling the propellants and minimizing mixing. In boosters with hypergolic propellants, the opposite happens— mixing is encouraged as these propellants burn on contact rather than mixing and exploding. In addition, the toxicity of hypergolic propellants means that it is desirable to burn them up as quickly as possible. The destruction system used on these boosters works by rupturing the common tank bulkhead allowing the oxidizer and fuel to come into immediate contact and burn. The stage engines are shut down just before the destruct charges are activated.

In addition to the RSO command, flight termination can also be initiated automatically. The range safety system tracks the launch vehicle and predicts its Instantaneous Impact Point (IIP), which is a real-time estimate of where the vehicle would land if the flight were terminated. Due to the high speed of the vehicle, the IIP can be several thousand kilometers downrange of the vehicle's current position. If for any reason the range safety personnel cannot verify that a vehicle is and will remain within the specified boundaries (e.g., if tracking systems fail), they will terminate the flight.

A key element of current range safety procedures is the definition of thresholds used to determine when a flight should be terminated. These thresholds are ultimately based on Impact Limit Lines (ILLs), which extend downrange from the launch site and define the area in which debris (from planned stage drops, vehicle explosions, or thrust termination) may land. Flight trajectories and ILLs are calculated and approved prior to launch to protect people and property. Destruct lines are defined within the ILLs to account for delays in operator response, uncertainties in vehicle breakup, wind, and other aerodynamic effects. If a vehicle's IIP reaches a destruct line, the flight is terminated. Wherever the vehicle passes over inhabited land masses before orbital insertion, "gates" (i.e., exits) are defined in the ILLs and destruct lines. The vehicle must pass through the gate or the flight will be terminated. The gates are perpendicular to the nominal trajectory, and the width of the gates takes into account tracking uncertainties and acceptable trajectory deviations.

1.7 Guidance and Control Actuators

The Scout launcher has all the possible attitude control actuators that can be accommodated on a rocket. During the first stage thrust, manoeuvring and attitude control are performed by movable wing tips located in the Section A of the launcher and by thrust vector control performed by jet vanes that deflect the flow exiting from the first stage nozzle. The second stage flight is controlled by cold gas jets located in the upper B interstage 1–2, and the same type of control is located in the upper C interstage 2–3 to guide and control the third stage engine. Finally the fourth stage is spin stabilised by a motor located in the upper D interstage 3–4.

The aerodynamic control of the first stage is provided by a set of cruciform wings located on the aft section of the launcher. The general assembly is in Fig. 1.18, which shows Section A connected to the first stage tank. The figure shows the jet vanes and the wing tip both encircled; the arrow indicates the common shaft. Each wing has a jet vane and a tip control mounted in a common torque tube, which allows the simultaneous rotation of both the surfaces. The wings are mounted on a cylindrical fuselage, the Base A, made of a semi-monocoque aluminium structure. The electrical wiring and the hydraulic plumbing to command the deflection of the wing tips and jet vanes are mounted on the inner walls of Base A in such a way that they to do not interfere with the attachment of Base A to the first stage engine [11]. Figure 1.19 shows the rear view of Base A: note the four jet vanes on the nozzle exhaust section.

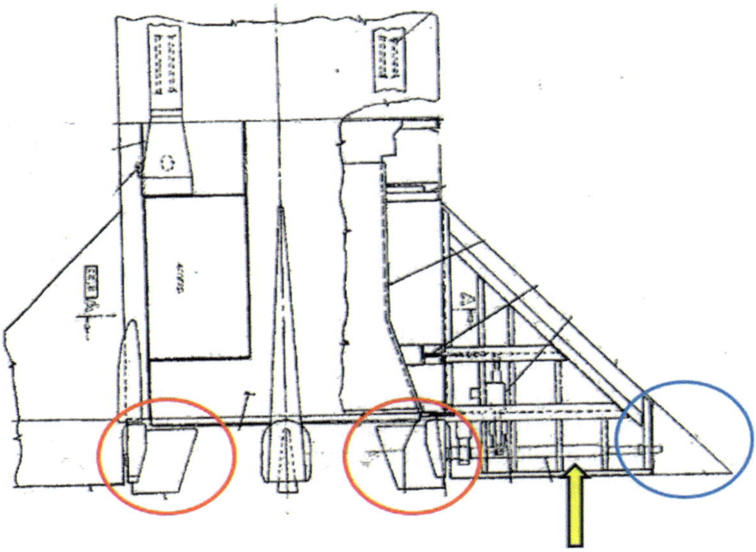

Fig. 1.18 The Scout base A, [5]

Fig. 1.19 Rear view of the Scout base A

The wing geometry is shown in Fig. 1.20, [11].

The upper B section contains the guidance and control (GC) system for the second stage: the piping and wiring are located on the inside wall of the upper B structure so as not to interfere with assembly of the section to the second stage. The GC system consists of hydrogen peroxide reaction thrusters, namely four 2220 N thrusters are used for the yaw and pitch motion and four thrusters of 180 N are used for the roll control. Ten fuel tanks are arranged in a circle and fixed inside the inner wall of the Upper B section. Figure 1.21 shows the assembly of the gas jet engines and the fuel

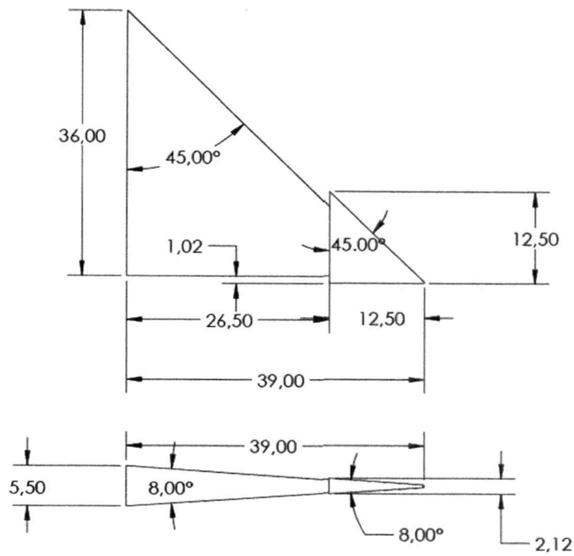

Fig. 1.20 Dimensions (cm) of the wings, [5]

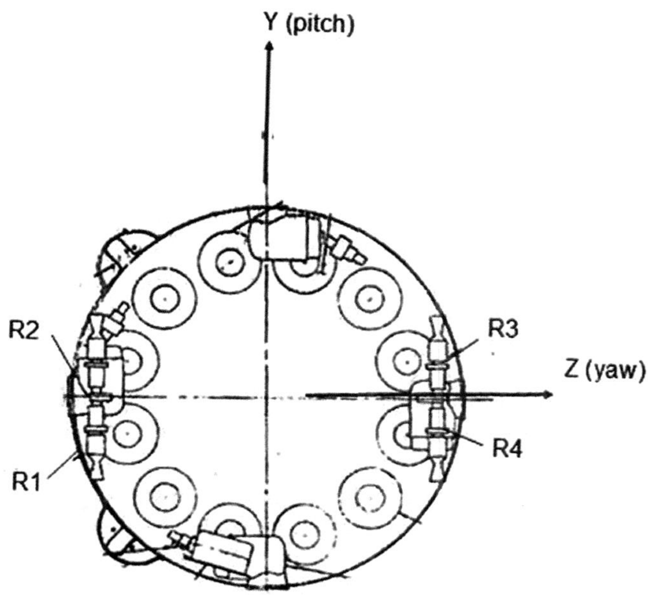

Fig. 1.21 The tanks and roll engine assembly on the inner wall of the upper B section, [5]

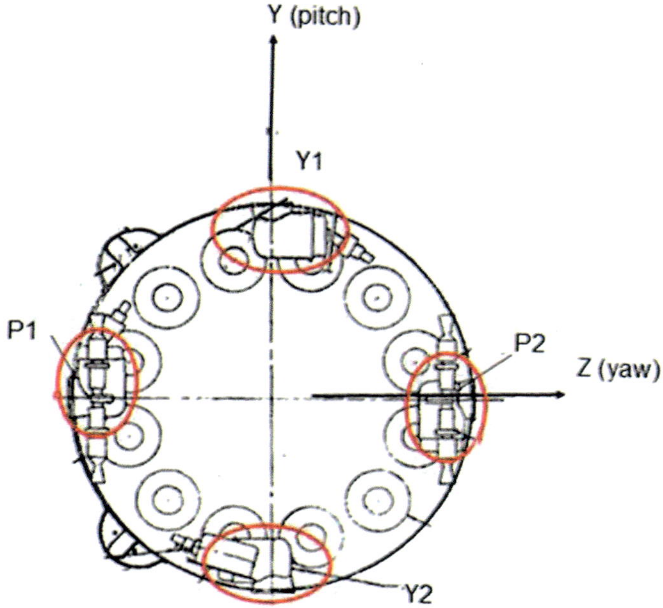

Fig. 1.22 The four pitch-yaw engines of thrust 2220 N are encircled by an ellipse, [5]

tanks with a looking forward view (the launcher nose cone is inside the page). The four roll engines are shown: positive roll rotation is provided by firing engines R1 and R3, negative roll is provided by the engines R2 and R4. The four pitch and yaw engines are underlined in Fig. 1.22. Ignition of P1 produces a positive pitch rotation, ignition of P2 a negative pitch rotation. Firing Y1 generates a positive yaw rotation, firing Y2 a negative yaw rotation [11]. Figure 1.23 shows a lateral view of the B sector. The roll jets and the tanks are indicated by a red ellipse, the second stage nozzle by a dashed blue line.

The third stage GC system is located in the inner wall of the upper C section. It consists of four jets for pitch and yaw of 250 N thrust each, and four roll jets, of 60 N thrust each. There are 2 tanks of H_2O_2 and two tanks of N_2 propellant. After the third stage burnout, during the coasting phase, the guidance system switches to other actuators. During this phase the 250 N jets are no longer used and pitch control is provided by two smaller jets of 9 N, feeded by the same tanks as the 250N system. Yaw control is mantained by the roll jets by a mixing logic that precludes the possibility of operating opposing jets simultaneously [11]. Figure 1.24 shows the third stage control system arrangement. The gas jets P1 and P2 are the 250 N pitch controller, the gas Jets p1 and p2 are the 9 N pitch controller during the coasting phase. The yaw is controlled by the 250 N gas jets Y1 and Y2 during the thrusting phase and by the 60 N roll gas jets during the coast phase. Namely R1 and R4

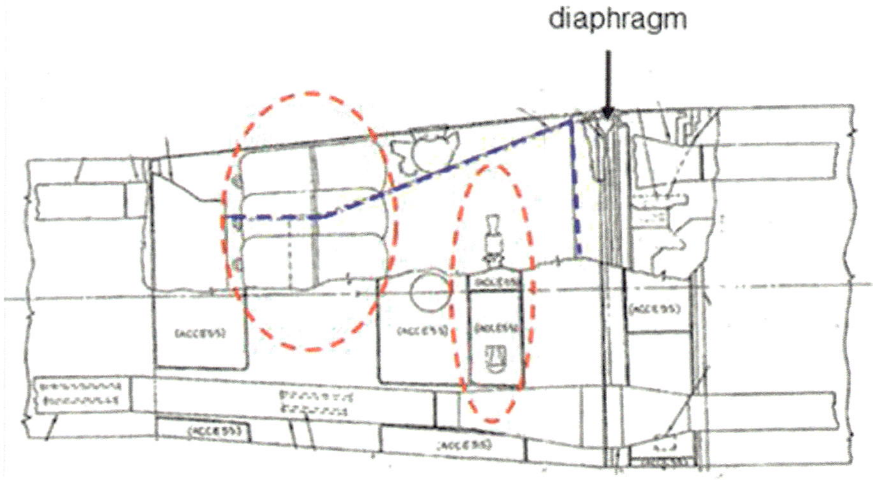

Fig. 1.23 Lateral view of the B section

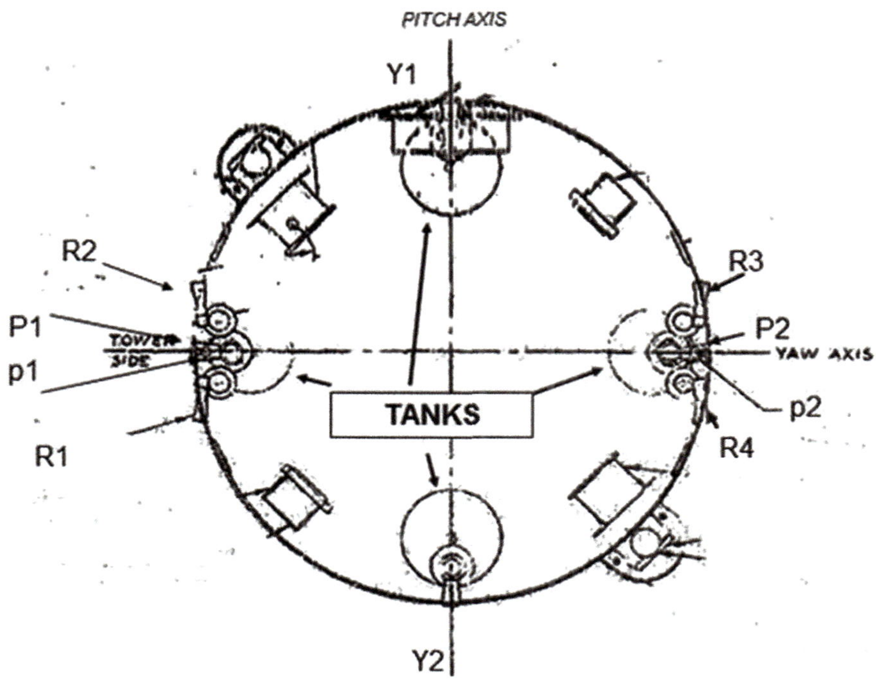

Fig. 1.24 Assembly of the third stage guidance and control system [5]

operated simultaneously produce positive yaw rotation, while R2 and R3 produce negative yaw rotation. The roll gas jets provide the roll control during the third stage thrust phase used in pairs R1-R3 (positive roll), R2-R4 (negative roll).

The fourth stage is spin stabilised by a spin bearing located in Section D. This consists of lower and upper sections. Lower D section attaches to the top of the third stage and supports the four stage spinning bearing. It also contains the guidance equipment. Upper D is attached to the four stage motor; its rear end is attached to the bearing ring which contains the spinning bearing and the separation diaphragm. Upper D also contains four spin engines which are actuated just before the separation between the third and fourth stages.

The four spin engines have a length of 12 cm, diameter equal to 4 cm and mass equal to 0.3 kg. The propellant is Archite high performance propellant producing a thrust of 190 N [10].

1.8 Rocket Aerodynamics

The aerodynamic properties of the Scout launcher are only relevant for the first and second stages, as at the second stage burnout the altitude is such that the air density is so low that negligible aerodynamic forces and moments are generated. This section presents Scout data for the main aerodynamic coefficients and the centre of pressure location as a function of Mach number and angle of attack [14].

Since the angle of attack α is limited during the atmospheric flight, the lift is considered linear in α, that is

$$L = \frac{1}{2}\rho\, V^2\, S\, C_{L_\alpha}\, \alpha$$

where the "stability derivative" C_{L_α} is in Fig. 1.25 for the full launcher at different Mach numbers. Note that the reference surface S has area of $0.5\,\mathrm{m}^2$.

The drag coefficient C_D is defined by the polar formula:

$$C_D = C_{D_0} + k_D\, (C_{L_\alpha}\, \alpha)^2 \tag{1.23}$$

where the drag at zero angle of attack C_{D_0} is in Fig. 1.26 and the coefficient k is in Fig. 1.27 for different Mach numbers.

The aerodynamic torque around the transversal axis (for instance pitch aerodynamic torque) is given by the formula

$$M = x_{sm}\, L \tag{1.24}$$

where the static margin x_{sm} is defined as the distance between the centre of mass position x_c and the centre of pressure position x_{cp} both counted from the nose [14]:

$$x_{sm} = x_c - x_{cp}$$

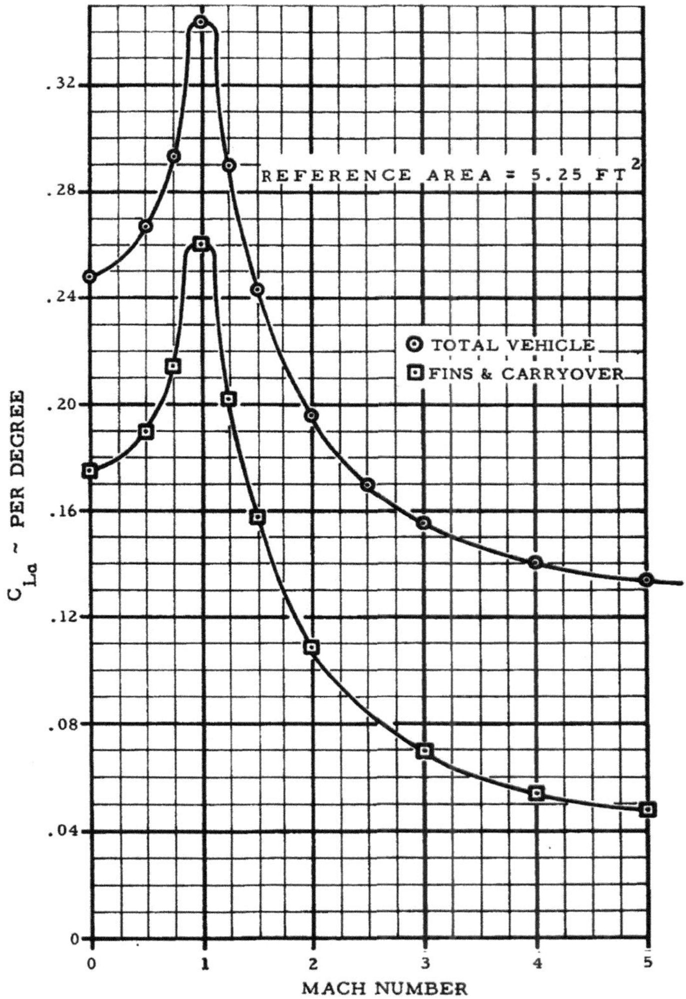

Fig. 1.25 The lift coefficient C_{L_α} for different Mach number: full Scout launcher [5]

Figure 1.28 shows the centre of pressure position (along the longitudinal axis) counted from the nose.

The atmospheric density ρ depends on the altitude and the effect of the Sun on the Earth's atmosphere. Average values of the atmospheric density with respect to the altitude are in the Table 1.13, which also gives the speed of sound needed to define the Mach number.

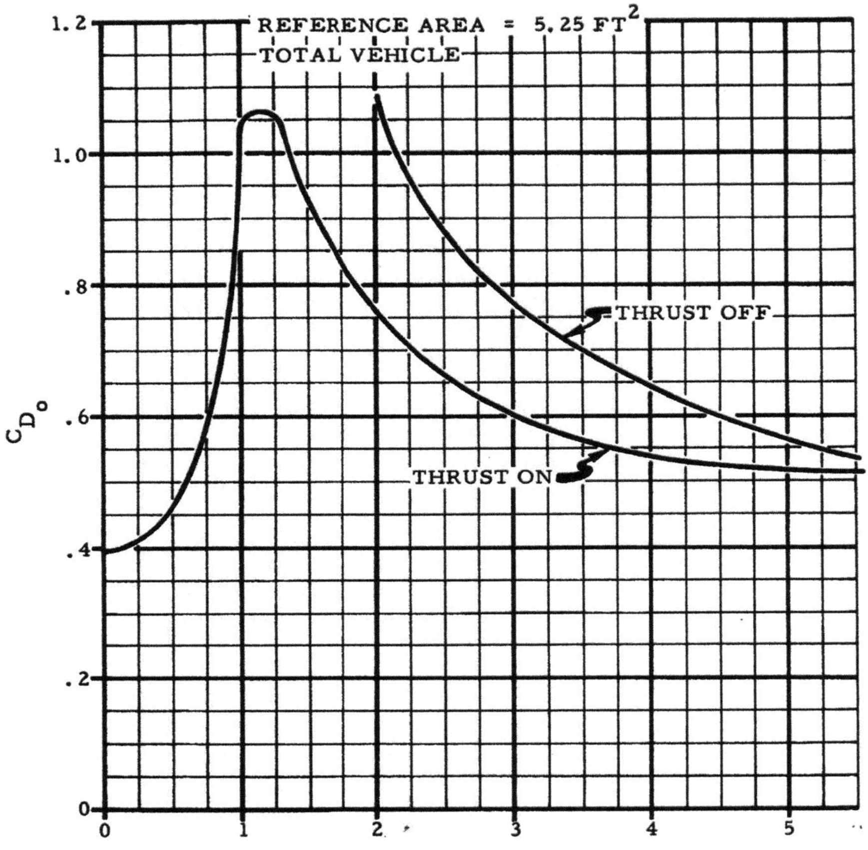

Fig. 1.26 The drag coefficient at zero lift C_{D_0} for different Mach number: full Scout launcher [5]

A numerical model of the atmosphere is the exponential model

$$\rho = \rho_k \, e^{-\beta_k \, (h_{k+1} - h_k)} \tag{1.25}$$

where the atmosphere is divided into N different layers at altitude h_k, $k = 1, \ldots N$. Altitudes h_k and corresponding values of atmospheric dentity ρ_k are taken from the first and second columns of Table 1.13 with N equal to 10. In Eq. (1.25) the scale factor β_k, $k = 1, N - 1$, holds in the layer $[h_k, h_{k+1}]$ and has the values:

$$\beta_k = [0.1090\,, 0.1568\,, 0.1549\,, 0.1602\,, 0.1476\,, 0.1364\,, 0.1194\,, 0.1254\,, 0.1567] \; 1/km$$

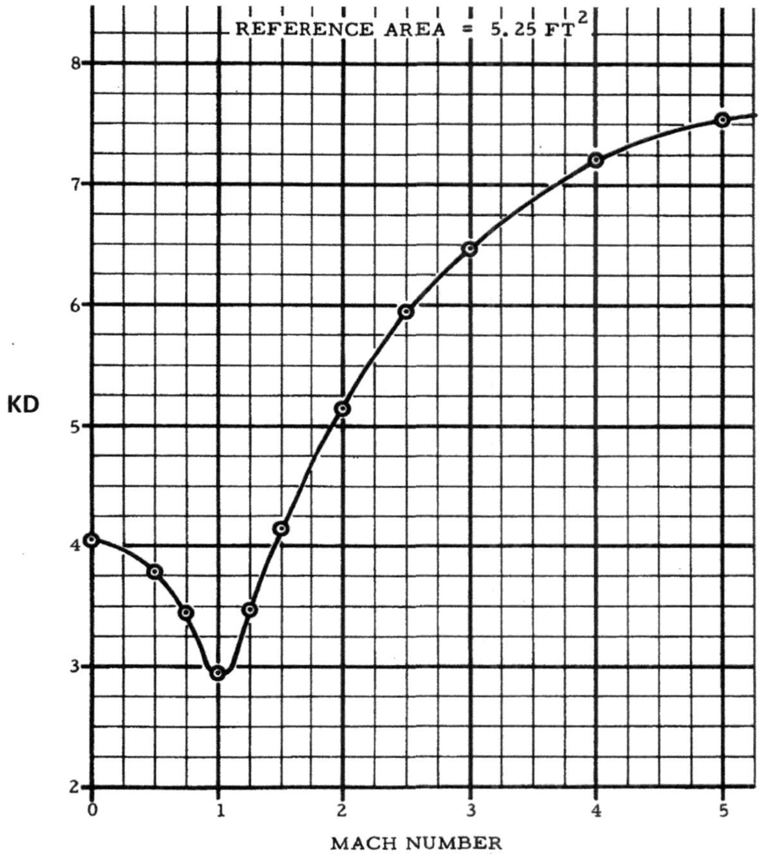

Fig. 1.27 The coefficient k_D of the polar formula (1.23) for different Mach number: full Scout launcher [5]

Figures 1.29, 1.30 show the lift coefficient and the centre of pressure location for subrocket 1, that is the full launcher minus the first stage.

1.9 The Launch Base and the Azimuth Equation

The launch site of a space rocket must be in a clear area to allow the safe fall of the first stages, which can impact thousands of kilometres away from the launch pad. The map in Fig. 1.31 shows that the current launch sites are either in front of the ocean or they have large inhabited ground areas around them. Even at these sites there may be restrictions on launch directions depending on range safety issues. For example,

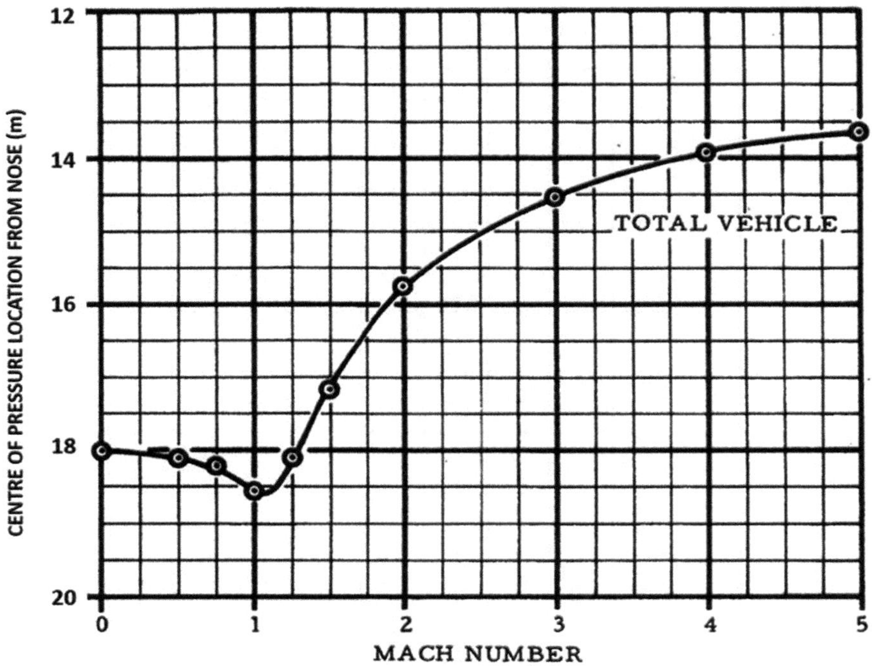

Fig. 1.28 The centre of pressure location x_{cp} counted from nose for different Mach number: full Scout launcher, [5]

Table 1.13 Atmospheric density (kg/m^3) and speed sound (m/s) for different altitudes (km)

Altitude (km)	Density kg/m^3	Sound speed (m/s)
0	1.22	340.3
10	0.41	299.5
16	0.16	295.1
26	3.4e-02	299.1
32	1.3e-02	303.0
40	3.99e-03	317.2
50	1.02e-03	330.0
60	3.09e-04	315.1
70	8.28e-05	297.1
80	1.84e-05	282.5

Fig. 1.32 shows that the San Marco launch base has launch directions within angles of 55 to 130 deg counted from North to avoid flying over populated areas.

These angles, called azimuth angles ψ, are determined by the target orbit plane with inclination i, and the launch pad latitude L: the relationship is:

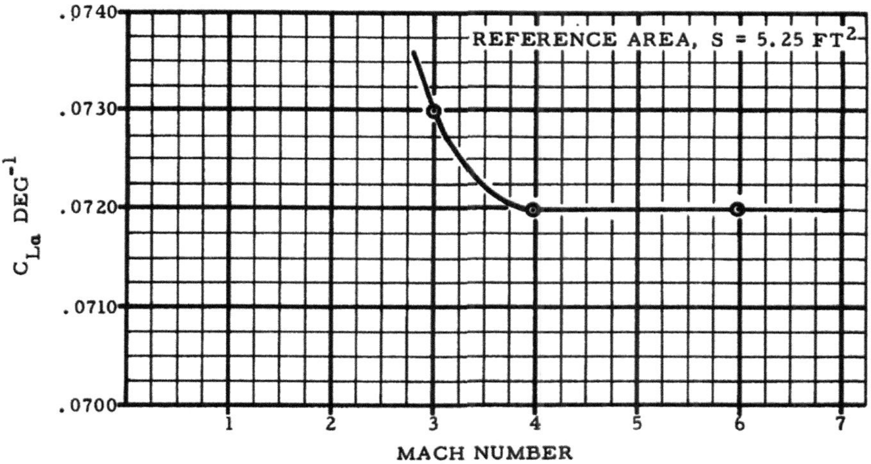

Fig. 1.29 The lift coefficient C_{L_α} for different Mach number: subrocket 1 [5]

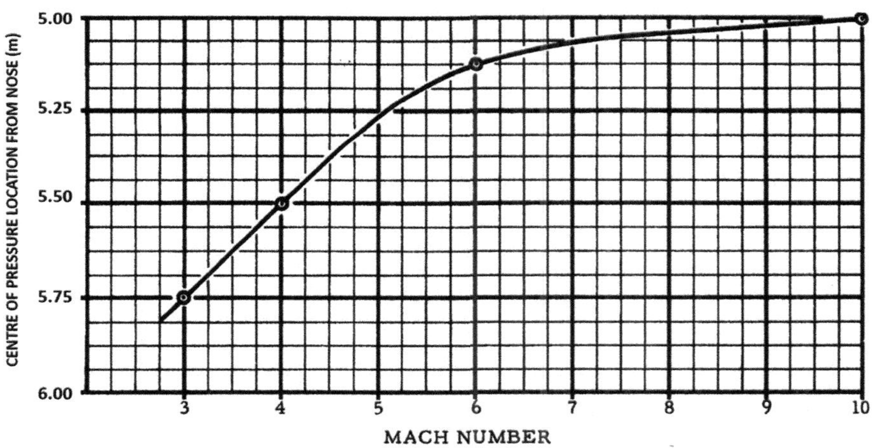

Fig. 1.30 The centre of pressure location x_{cp} counted from nose for different Mach number: subrocket 1 [5]

The Launch Base Equation

$$\sin\psi = \frac{\cos i}{\cos L} \tag{1.26}$$

From this equation it follows:

$$\sin\psi = \frac{\cos i}{\cos L} < 1 \longrightarrow \cos i < \cos L \longrightarrow i > L$$

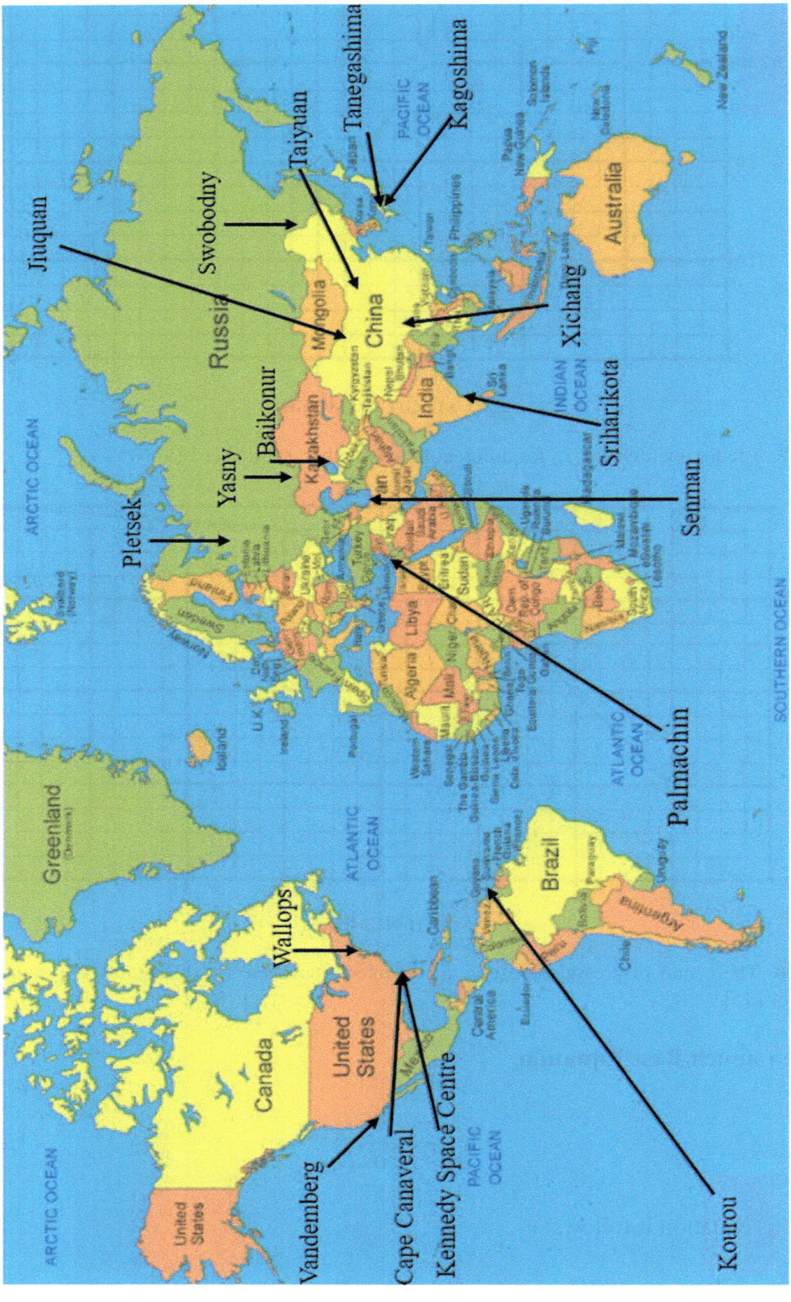

Fig. 1.31 Operative launch bases in the World

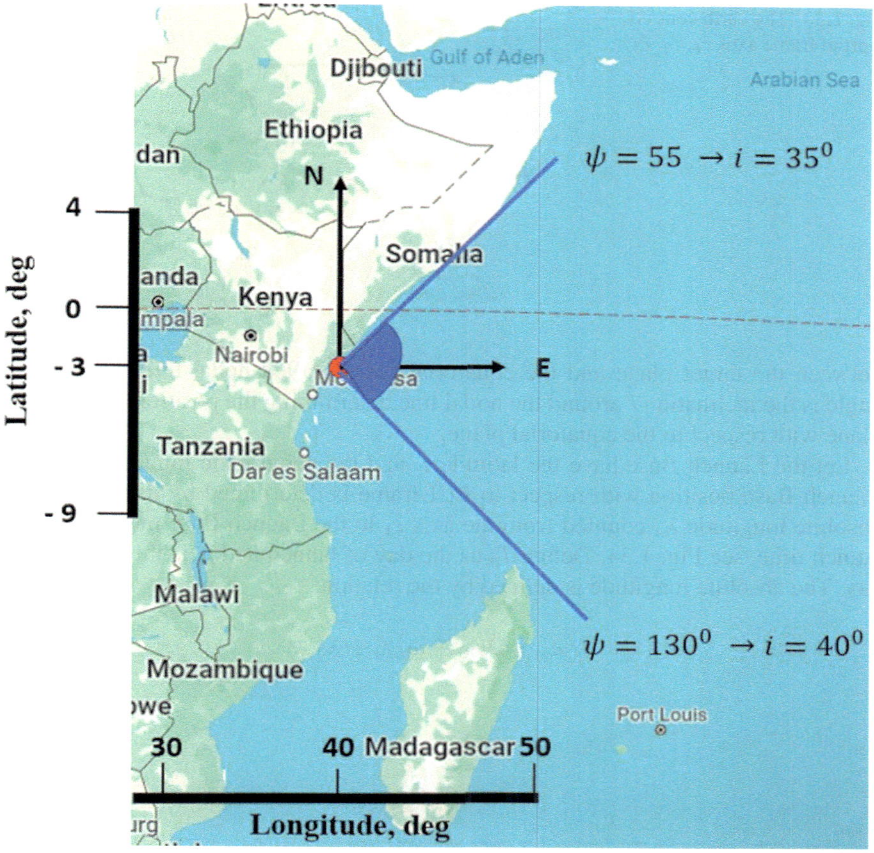

Fig. 1.32 The allowed azimuth angle for launches from San Marco platform

i.e. it is only possible to select an azimuth angle ψ if the latitude L of the launch base is less than the inclination i of the required orbit plane. If this is not the case, an expensive out of plane manoeuvre is required to reach the target orbital plane. For this reason launch sites around the world are located as close to the equator as possible, since any orbital inclination can be achieved from an equatorial launch base (if there are no other constraints).

The proof of the Launch Base Eq. (1.26) is now given.

The inclination of the plane of the target orbit is an angle determined by the Earth Centred Inertial (ECI) which is defined as it follows: the origin is at the centre of the Earth, the axes $\hat{c}_1$, $\hat{c}_2$ are in the equatorial plane and the axis $\hat{c}_3$ is perpendicular to the equator and faces north. The direction $\hat{c}_1$ is fixed at the intersection of the ecliptic and equatorial planes, see Fig. 1.33.

The plane of the target orbit is identified by two angles. The first is the Right Ascension Ascending Node (RAAN) angle Ω from $\hat{c}_1$ to the line of intersection

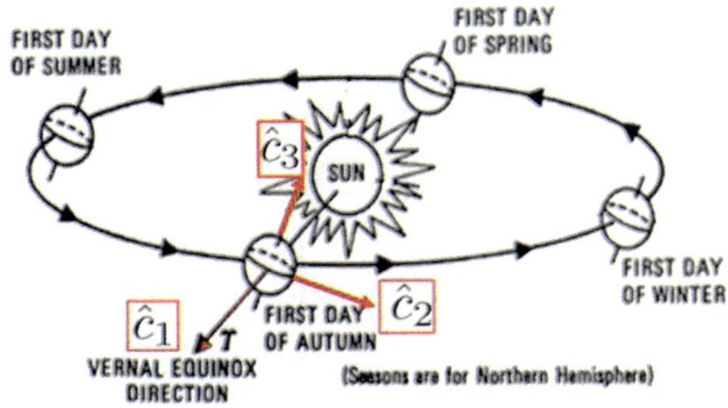

Fig. 1.33 The earth centred inertial frame axes $\hat{c}_1, \hat{c}_2, \hat{c}_3$

between the target plane and the equatorial plane, called nodal line. The second angle is the inclination i around the nodal line determining the position of the target plane with respect to the equatorial plane.

Let the Launch Base have the latitude L and the geographic longitude λ_g. The Launch Base position with respect to ECI frame is determined by its latitude and absolute longitude λ_a counted from the axis $\hat{c}_1$ to the Launch Base meridian at the launch time, see Fig. 1.34. Define T_0 as the day of launch and t_0 is the time of that day. The absolute longitude is defined by the relation:

$$\lambda_{a_0} = \lambda_{G0} + \omega_E t_0 + \lambda_g$$

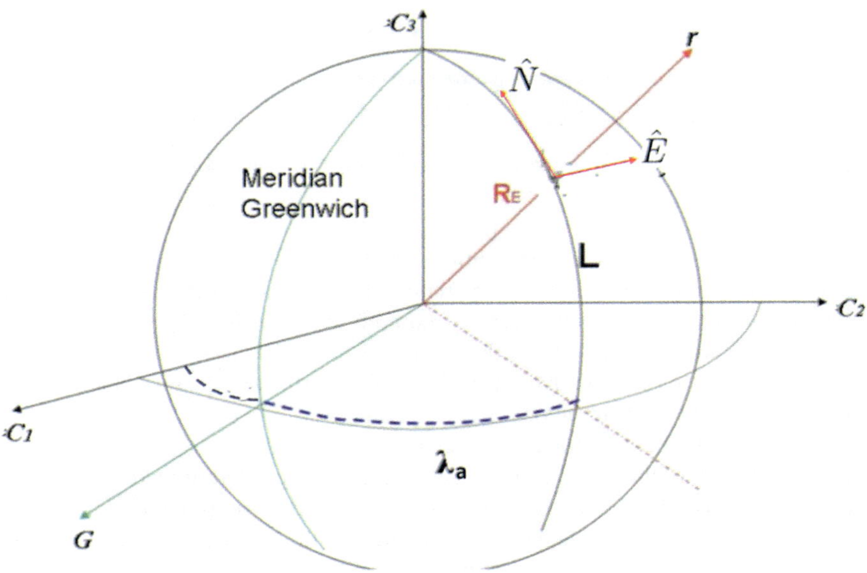

Fig. 1.34 The launch site reference frame $(\hat{r}_0, \hat{E}_0, \hat{N}_0)$, the absolute longitude λ_{a_0} and latitude L

where λ_{G0} is the Greenwich absolute longitude at a reference day T_0. If τ_0 is the Julian date of the date T_0, then the corresponding Greenwich absolute longitude is in degrees:

$$\lambda_{G0} = mod(99.6909832 + 36000.7689\, \tau_0 + 0.00038708\, \tau_0^2,\ 360)$$

The Launch Base Inertial (LBI) frame $(\hat{r}_0, \hat{E}_0, \hat{N}_0)$ has the radial direction $\hat{r}_0$ and the east- north directions of the launch base at the fixed time of launch.

This reference frame is obtained from the $(\hat{c}_1, \hat{c}_2, \hat{c}_3)$ frame by a 3-2 sequence of rotations of absolute longitude λ_{a_0} and latitude L.

$$\begin{bmatrix} \hat{r}_0 \\ \hat{E}_0 \\ \hat{N}_0 \end{bmatrix} = \begin{bmatrix} c_L & 0 & s_L \\ 0 & 1 & 0 \\ -s_L & 0 & c_L \end{bmatrix} \begin{bmatrix} c_{\lambda_{a_0}} & s_{\lambda_{a_0}} & 0 \\ -s_{\lambda_{a_0}} & c_{\lambda_{a_0}} & 0 \\ 0 & 0 & 1 \end{bmatrix} \begin{bmatrix} \hat{c}_1 \\ \hat{c}_2 \\ \hat{c}_3 \end{bmatrix}$$

The inverse rotation:

$$\begin{bmatrix} \hat{c}_1 \\ \hat{c}_2 \\ \hat{c}_3 \end{bmatrix} = \begin{bmatrix} c_{\lambda_{a_0}} & -s_{\lambda_{a_0}} & 0 \\ s_{\lambda_{a_0}} & c_{\lambda_{a_0}} & 0 \\ 0 & 0 & 1 \end{bmatrix} \begin{bmatrix} c_L & 0 & -s_L \\ 0 & 1 & 0 \\ s_L & 0 & c_L \end{bmatrix} \begin{bmatrix} \hat{r}_0 \\ \hat{E}_0 \\ \hat{N}_0 \end{bmatrix}$$

gives

$$\hat{c}_3 = \sin L \hat{r}_0 + \cos L \hat{N}_0 \tag{1.27}$$

From Fig. 1.34 il follows that the normal to the target plane $\hat{h}$ verifies:

$$\hat{h} \cdot \hat{c}_3 = \cos i$$

$$\hat{h} \cdot \hat{N}_0 = \sin \psi$$

From the above equations and from (1.27) one has:

$$\cos i = \hat{h}\hat{c}_3 = \cos L\ (\hat{h}\, \hat{N}_0) = \cos L\ \sin \psi$$

which is equivalent to the Launch Base Eq. (1.26).

From the Eq. (1.26), the azimuth angle is:

$$\psi = \arcsin\left(\frac{\cos i}{\cos L}\right)$$

The azimuth angle selects the required plane from all the possible planes passing through the radial direction $\hat{r}_0$, and it would be sufficient to direct the thrust on this plane to achieve the required orbital inclination. However this is not quite true because of the Earth's rotation. In fact the launch pad, and therefore the rocket, has an inertial velocity towards the east:

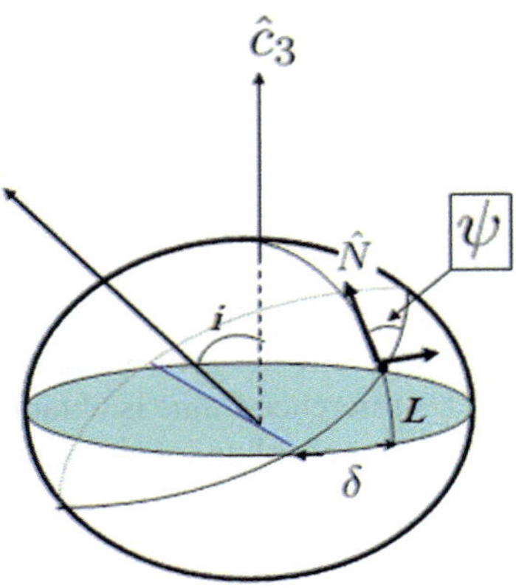

Fig. 1.35 Azimuth and δ angles

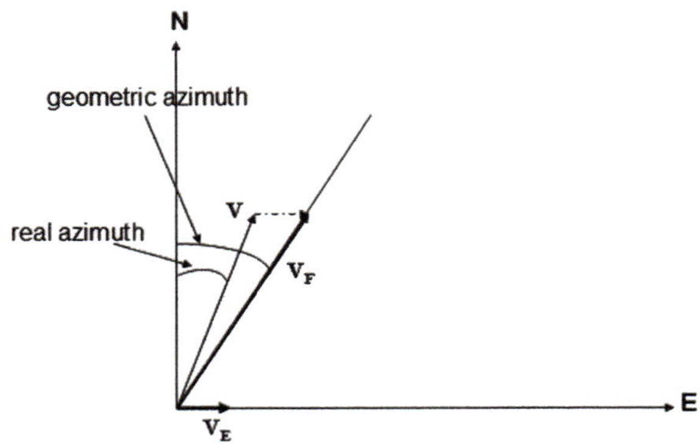

Fig. 1.36 The real azimuth angle

$$\mathbf{v}_E = \boldsymbol{\omega}_E \times \mathbf{R}_E = \omega_E \, R_E \cos L \, \hat{E}. \tag{1.28}$$

(ω_E Earth angular velocity: $7.29\,e^{-5}\,\frac{\text{rad}}{\text{s}}$) depending on the latitude of the launch base; the maximum velocity occurs when the launch base is equatorial: $v_E^{eq} = 463\,\frac{\text{m}}{\text{s}}$.

The condition is shown in Fig. 1.36: we see the trace of the target plane, the geometric azimuth angle ψ and the velocity towards east that the launcher has on the launch pad. To compensate for the drift towards the east, the real azimuth angle must be smaller than the geometric azimuth angle given by the Eq. 1.26. In this way the current velocity $\mathbf{V}$ added to the eastward drift $\mathbf{V}_E$ gives the final velocity $\mathbf{V}_f$ which lies on the required plane.

To evaluate the real azimuth angle and the relative velocity needed to reach the orbital velocity $V_f = \sqrt{\frac{\mu}{r_f}}$, let assume that the launch site is at latitude L and the inclination of the target orbit is i. Then the azimuth angle ψ verifies: $\sin\psi = \frac{\cos i}{\cos L}$ and the final velocity has east/north components:

$$\mathbf{V}_f = V_f \cos\phi \, \hat{N} + V_f \, \sin\phi \, \hat{E}$$

The relative final velocity is $\mathbf{V}_{rel_f} = \mathbf{V}_f - \mathbf{V}_E$, where $\mathbf{V}_E = \omega_E \, R_E \, \cos L \, \hat{E}$. Then $\mathbf{V}_{rel_f}$ has east/north components:

$$\mathbf{V}_{rel_f} = V_f \cos\psi \hat{N} + (V_f \, \sin\psi - \omega_E \, R_E \, \cos L) \, \hat{E}$$

Therefore the real azimuth angle ψ_r is given by:

$$\sin\psi_r = \frac{V_f \, \sin\psi - \omega_E \, R_E \, \cos L}{V_r}$$

$$\cos\psi_r = \frac{V_f \, \cos\psi}{V_r}$$

where

$$V_{rel_f} = \sqrt{V_f^2 + V_E^2 - 2V_f \, V_E \frac{cos i}{\cos L}} \tag{1.29}$$

Note that for equatorial launch bases and equatorial orbits the relative velocity is minimum: $V_{rel_f} = V_f - V_E = V_f - \omega_E \, R_E$. Minimum relative velocity is required to achieve orbits with inclination equal to the latitude of the launch base ($i = L$): $V_{rel_f} = V_f - \omega_E \, R_E \, \cos L$. Therefore in these cases the variation of the velocity to be provided by the launcher is smaller, so more payload mass can be placed in these orbits. On the other hand for polar orbits the maximum relative velocity is required: $V_{rel_f} = \sqrt{V_f^2 + V_E^2}$ and less payload can be put into orbit.

1.9.1 Launch Window

To achieve the target orbit inclination at the required RAAN (Right Ascension of Ascending Node) Ω, it is necessary to synchronise the launch time so that after the vertical trajectory the launcher reaches the target plane at the required Ω. Considering an equatorial launch site, the absolute longitude of the launch site must be equal to the RAAN of the target orbit:

$$\lambda_a = \Omega$$

and the azimuth angle $\psi = \frac{\pi}{2} - i$. As it can be seen from the Fig. 1.37 it is necessary to wait for the right time t^* so that

$$\lambda_a(t^*) = \theta_{g0} + \lambda_g + \omega_E \, t^* = \Omega \tag{1.30}$$

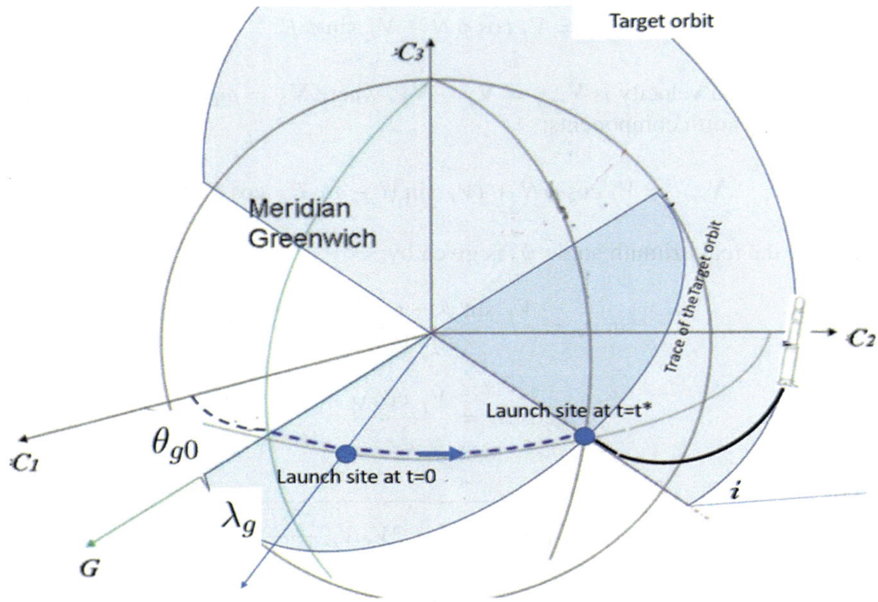

Fig. 1.37 The launch time t^* for the equatorial launch site

where: t^* is the time of the launch date such that the launch meridian intersects the nodal line of the target orbital plane, θ_{g0} is the absolute longitude of the Greenwich meridian at time $0^h\, 0^{min}\, 0^s$ of the launch date,[1] and λ_g is the geographic longitude (that is counted from Greenwich) of the launch site, see Fig. 1.37.

The above formula defines the launch time t^* at a given date. Of course there is another possibility of launch at the descending node, after exactly 12 h:

$$\lambda_a(t^* + 12h) = \theta_{g0} + \omega_E\,(t^* + 12h) + \lambda_g = \Omega + \pi$$

Now consider a launch site at latitude $L \neq 0$. Figure 1.35 shows that the absolute longitude of the launch site must be equal to the sum

$$\lambda_a = \Omega + \delta$$

where δ is the launch site longitude computed with respect to the nodal line.

[1] For example θ_{g0} =249.7 deg at time 0 of the date 1st June 2025.

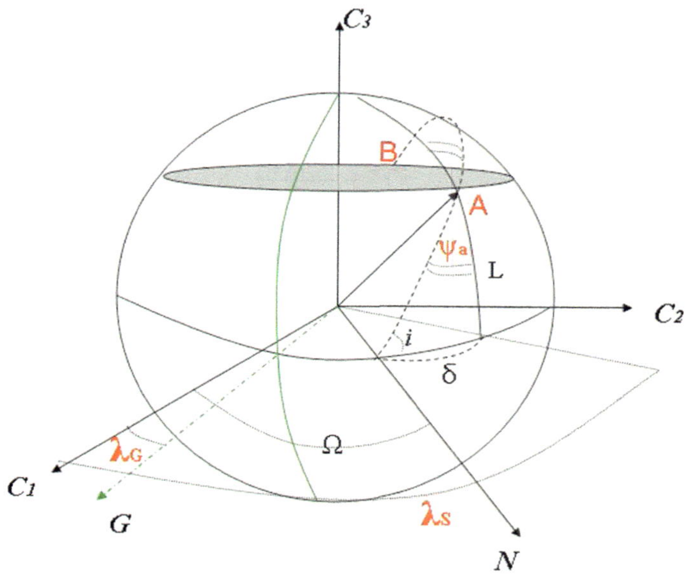

Fig. 1.38 The two launch opportunities A and B

Looking at the spherical triangle defined by the inner angles i, ψ, $\frac{\pi}{2}$ and the outer angles δ and L, the spherical law of sines gives:

$$\frac{\sin \psi_a}{\sin \delta} = \frac{\sin\ i}{\sin\ L}$$

then

$$\sin \delta = \sin \psi_a \frac{\sin\ L}{\sin\ i} = \frac{\cos i}{\cos L} \frac{\sin\ i}{\sin\ L} = \frac{\tan L}{\tan i}$$

It is apparent that there are two launch opportunities during a day since during a complete rotation of the Earth the launch Base intersects the trace of the target plane in two points A and B, see Fig.1.38.

In the singular case $L \equiv i$, (orbit inclination equal to the launch base latitude) there is just a launch opportunity in a day (A = B).

To calculate the launch time t^*, let us consider the equation for the absolute longitude:

$$\lambda_a(t^*) = \theta g0 + \omega_E\, t^* + \lambda_g = \Omega + \delta$$

The second launch opportunity is at time t^{**}:

$$\lambda_a(t^{**}) = \theta g0 + \omega_E\, t^{**} + \lambda_g = \Omega + \pi - \delta$$

The time elapsed between the two launch opportunities is obtained by subtracting the two:

$$\omega_E(t^{**} - t^*) = \pi - 2\delta$$

Then

$$t^{**} = t^* + \frac{\pi - 2\delta}{\omega_E}$$

As noted above there is only one launch opportunity if $i = L$, in which case $\delta = \frac{\pi}{2}$ and $t^{**} = t^*$.

1.9.2 Alternative Launch Sites

In addition to ground based launch stations, air-launched and sea-launched systems are also being considered for space missions because of their flexibility: the space launcher can be released at a convenient point of the Earth, bypassing the restrictions on the orbital inclinations that can be achieved. These options are particularly attractive to nations that have no capability to launch safely from the ground. The disadvantage is that payloads of reduced mass can be placed in orbit, but this issue is becoming less critical with the success of the microsatellite format.

The air-launch concept was developed into commercial service by the Pegasus project which ran from 1993 to 2021 [15]. The launcher was a three stage solid propellant rocket with an optional fourth stage using liquid propellant (hydrazine). The maximum payload was 400 kg and the aircraft carrier of the rocket was initially a Boing B-52 and later a modified Lookheed L-1011.

The sea-launch option is being developed by China from 2019 using various marine platforms and the Long March 11, a four stage solid fuel rocket [16].

The first sea-based launch took place in 1966 from the San Marco platform. The San Marco project was developed by the Centro Ricerche Aerospaziali of the School of Aerospace Engineering of Rome, under the direction of its Dean, Professor Luigi Broglio. The project included the manufacture of the San Marco satellites and the launch operations using the Scout launcher. The first San Marco satellite was launched by Scout from Wallop Station in 1964 and San Marco B was launched from the San Marco platform in 1966 [17].

The San Marco Equatorial Mobile Range actually consists of a launch platform (S. Marco) a control platform (S. Rita) located at about three miles off the the coast of Malindi, Kenya and a ground based logistics centre, see Figs. 1.39 and 1.40.

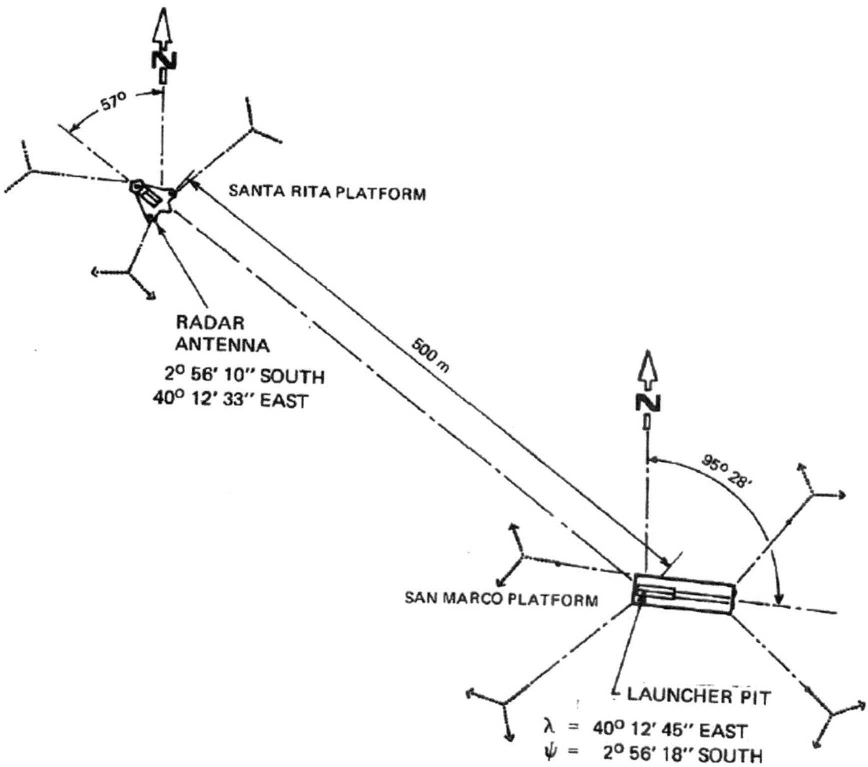

Fig. 1.39 The San Marco equatorial base geographical location [18]

The San Marco platform is a 3000 ton steel barge equipped with a lifting system consisting of 20 pneumatically operated cylindrical legs. Each leg is 1.8 m in diameter and 30.5 m long. The planar projection of the barge is 27.5 m wide and 91.5 m long, the depth of the hull is 4 m.

The arrangement of the platform is in Fig. 1.41 with a planar view in the upper part of the figure and a vertical profile in the lower part.

1.9.3 Launch Operations

The rocket motors, interstages and igniters are shipped by boat and they are loaded onto the San Marco platform by the platform crane, which lifts the containers on the platform deck, see Fig. 1.42.

Fig. 1.40 The San Marco and Santa Rita platforms

The rocket engines are placed in the engine store room of Fig. 1.41. The following launch operations are performed [20]:

1. Preparation of the transporter, a truck for moving and integrating the stages.
2. First stage build up:
   (i) removal of the stage from shipping container
   (ii) removal of the igniter from the stage shipping container and storage in the pyro-room of Fig. 1.41
   (iii) loading of the stage on the transporter
   (iv) horizontal alignment of the first stage with the launch tower
   (v) installation of the of supports for the external tunnels (cables, safe & arm and destruction system) along the longitudinal axis of the stage
   (vi) installation of the auto destruction module and inert safe and arm
   (vii) installation of the head cap pressure transducer
   (viii) integration on the transporter of Base A with the first stage
   (xi) installation of the wire bundle using clamps and installation of the tunnel cover
   (x) leak test of the reaction control system

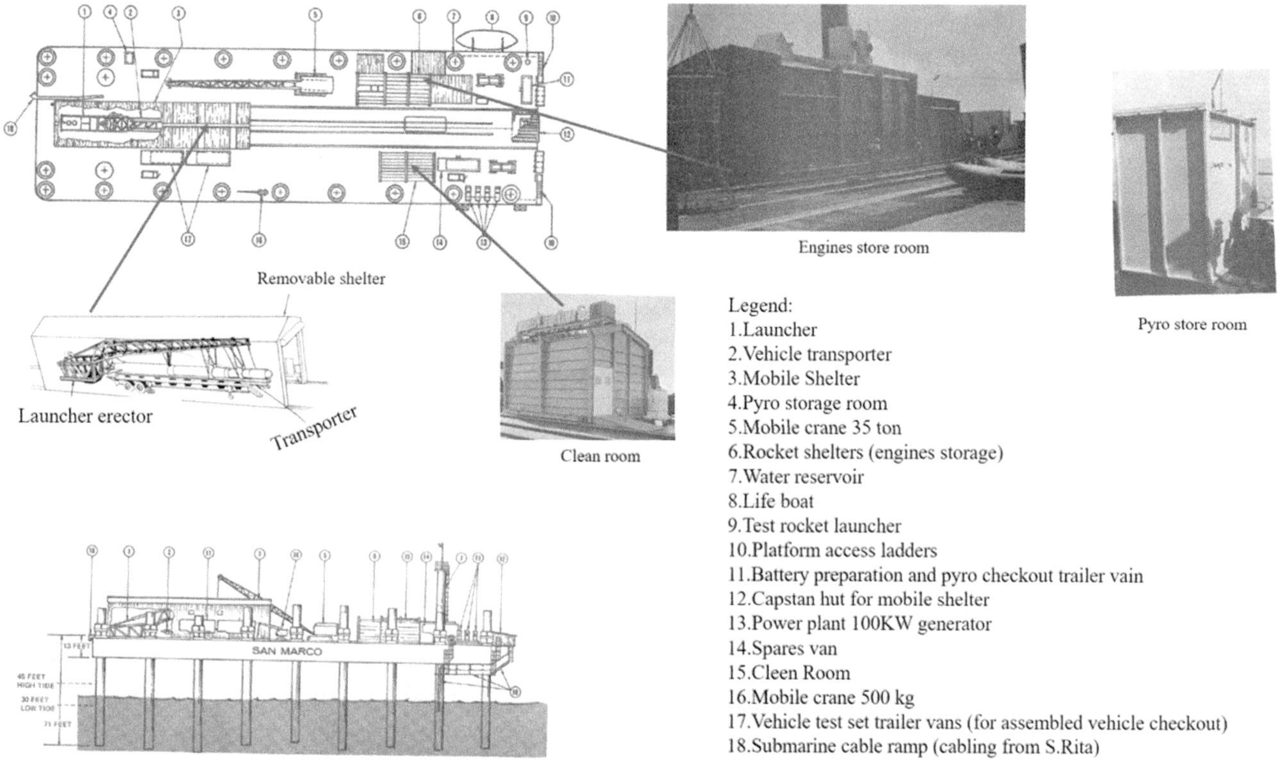

Fig. 1.41 The assembly of the San Marco platform [19]

Fig. 1.42 Loading the Scout rocket from the Tennessee boat onto the San Marco Platform

3. Second and subsequent stages built up
   the same steps (i)–(ii) and (v)–(vii) are carried out; step (viii) is replaced by the integration of the upper and lower parts of the interstage, after aligning the stage with respect to the previous stage by means of vertical movements of the cradles located on the transporter and, if necessary, rotation around the longitudinal axis.
4. Inserting of the entire transponder inside the mobile shelter
5. Attachment of the integrated rocket to the launch tower
6. System tests
7. Removal of the transporter
8. Movement of the mobile tunnel to clear the launch tower
9. Launch tower erected
10. Crew moved to the Santa Rita Platform.

Figure 1.43 shows the Scout second stage being removed from the shipping container. The cover of the container is removed and and motor tiedown bands are released. The igniter is also removed from the motor shipping container and moved to the Pyro-Room. The stage is then moved by a mobile hoist to be loaded over the transporter.

The transporter has four pairs of cradles to host the stages.

Figure 1.44 highlights the cradles of the first stage and the alignment conditions between the first stage aft and the lower cradles. These can be moved horizontally and vertically and also allow rotation around the longitudinal axis of the transporter.

Each stage is carefully placed on the cradles of the transporter. Figure 1.45 shows the third stage being placed on the transporter before entering inside the mobile tunnel for alignment with the horizontal launch tower.

The integration between the stages takes place on the transporter.

The wire bundles for the Safe&Arm and the destruction command are also integrated on each stage.

Figure 1.46 shows the preparation of the cable connections between the first and second stages.

Figure 1.47 shows the tunnel support installation for the destruction system.

Figure 1.48 shows the electrical wire bundle deployed along the launcher using wire clamps [13].

Figure 1.49 shows the transporter and the integrated launcher inside the movable tunnel. The next operation is to load the integrated rocket on the launch tower, then remove the transporter and the movable tunnel, see Fig. 1.50. The launch tower can be erected and the operations on the San Marco platform are complete.

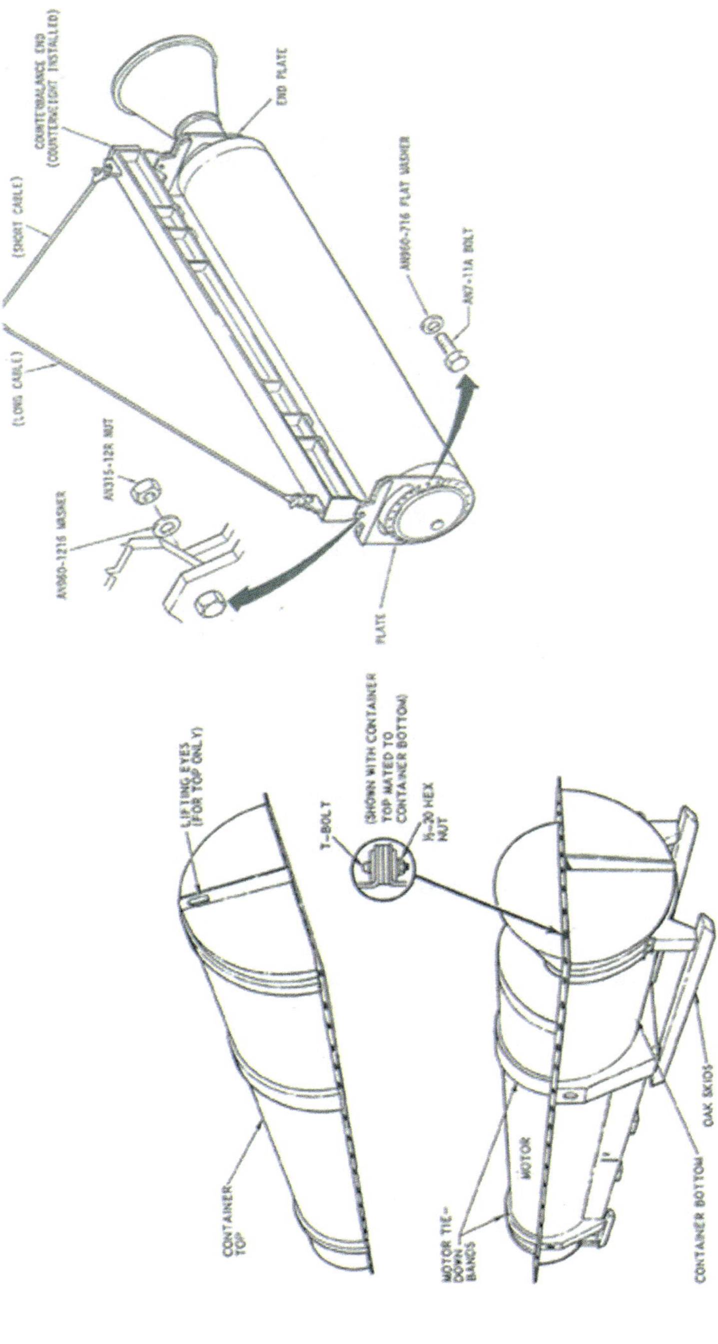

Fig. 1.43 Transfer from shipping container to the platform deck [5]

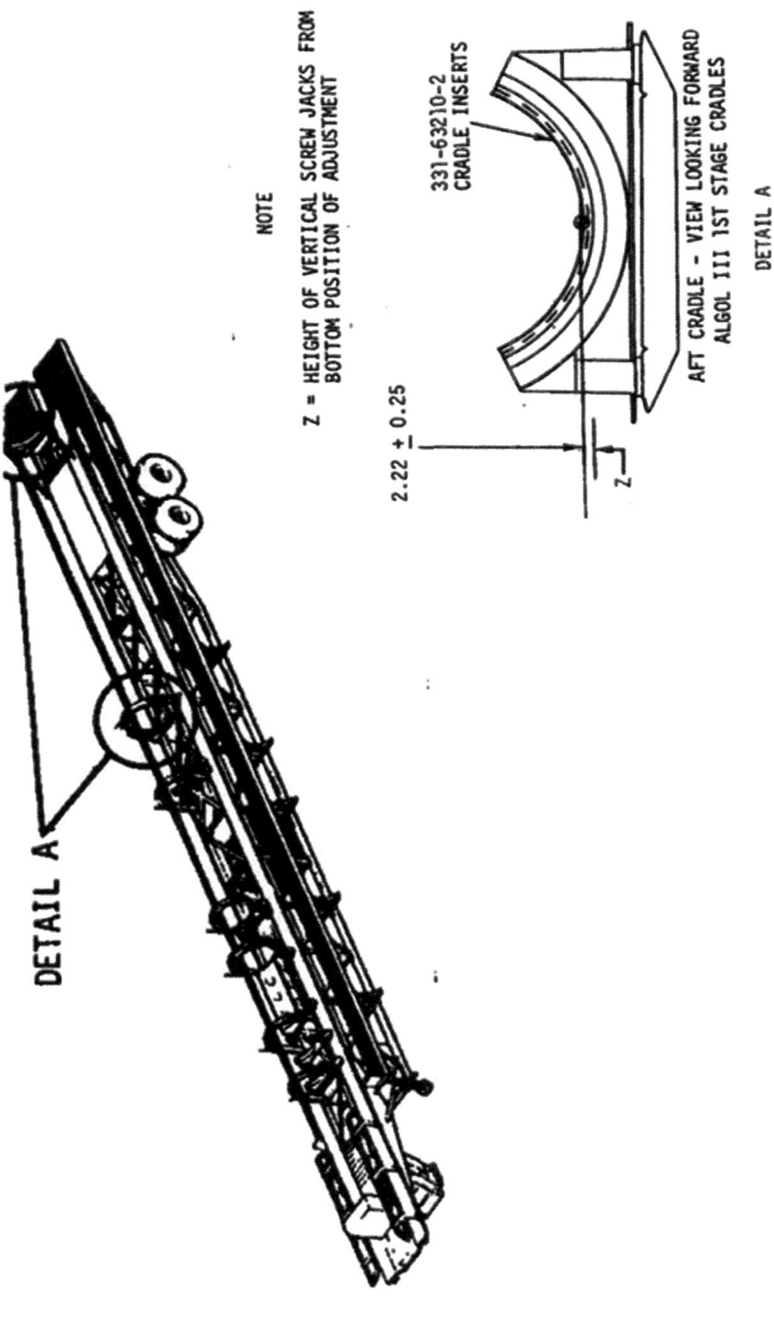

Fig. 1.44 Cradles of the first stage: location on the transporter and alignment conditions [5]

Fig. 1.45 Scout third stage loading onto the transporter

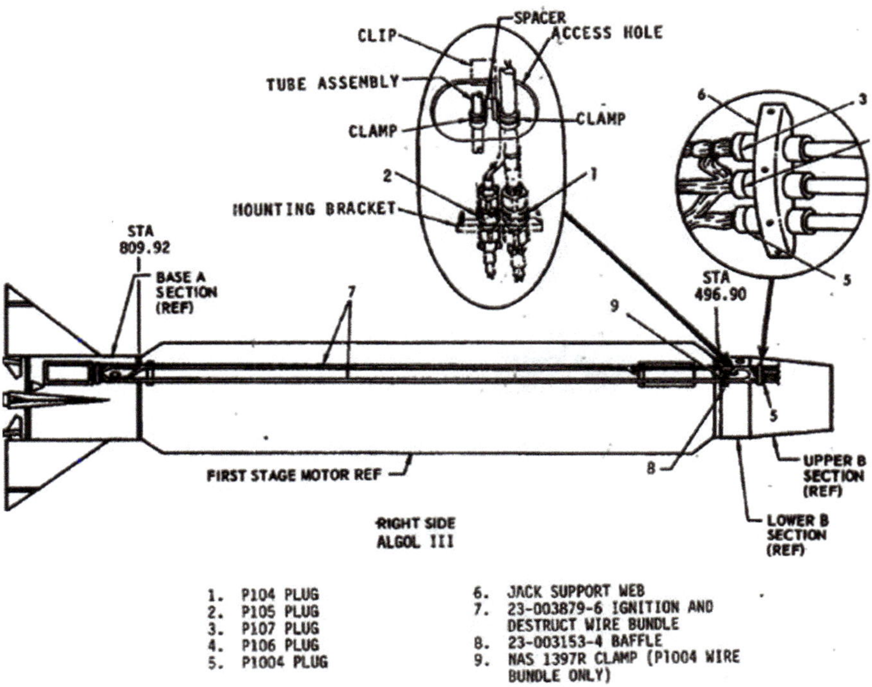

Fig. 1.46 Cabling the first and second stage in the B section interstage [13]

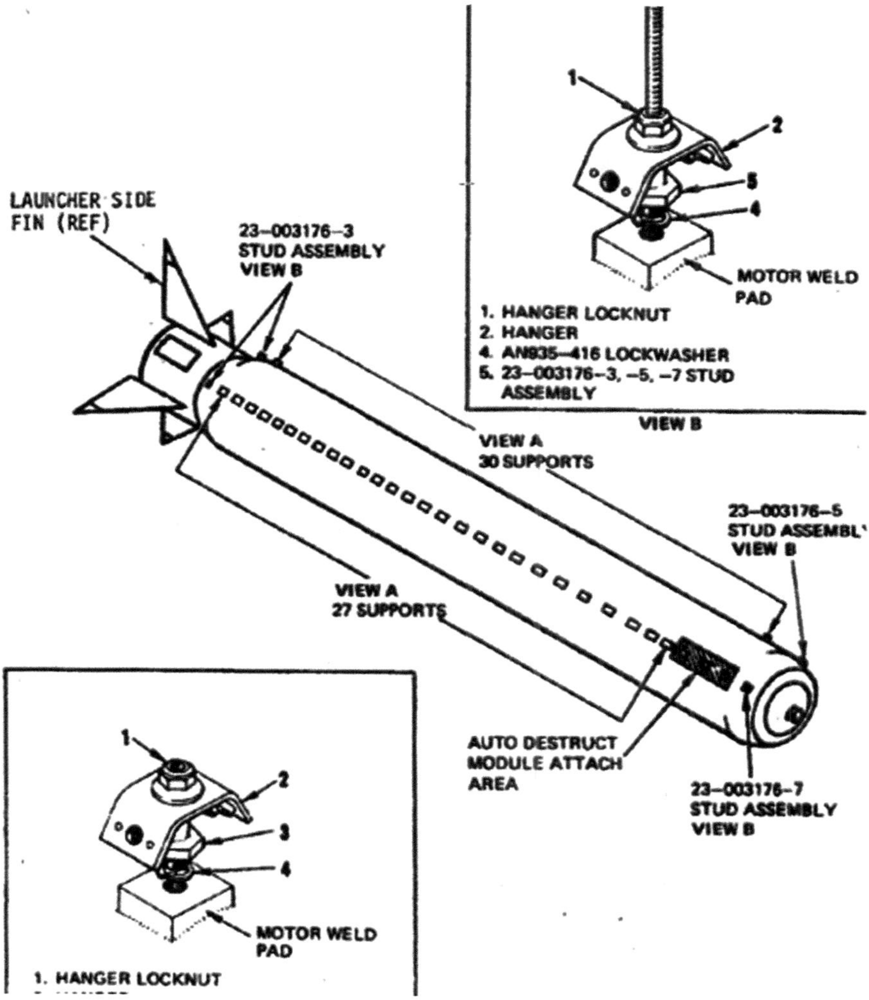

Fig. 1.47 The supports of the tunnel for the auto destruction system of the first stage [5]

1.9.4 Launch Control Infrastrucutres

The launch control was carried out from a second platform called Santa Rita. This is a drilled type three leg platform of shape of an isosceles triangle of base 35 m an height 36.5 m, see Fig. 1.51.

Fig. 1.48 The wire bundle deployed along the first stage and fixed on supports by clamps

The legs are electronically operated; the platform houses the blockhouse equipment, launch support and telecommunications and telemetry equipment. Other smaller decks contains administrative compartments and cruising quarters. The Santa Rita assembly is in Fig. 1.52.

Two small auxiliary platforms are located near Santa Rita: these are four-legged not floatable platforms. The larger (10x10 meters) supports the radar and optical systems for the space rocket tracking. The smaller one (8x8 meters) supports 6 diesel power generators producing 100 KW each.

The tasks of the Santa Rita platform are:

- Provide meteorological data close to the lift off procedure.
- Control the lift-off operations
- Tracking of the launcher and activating the rocket destruction system, if necessary.

Fig. 1.49 The transporter with the integrated rocket and the launch tower inside the movable tunnel

Fig. 1.50 The launch tower with the Scout launcher ready for erasing

Fig. 1.51 The Santa Rita platform

Measurements of wind, temperature, pressure, humidity and wave heights are collected up to few minutes before the lift off. The meteorological support facilities hosted in the Santa Rita platform are: (i) anemometers, (ii) radiosonde carried aloft by gas filled balloon with pressure, temperature and humidity sensors, (iii) rawin set (AN/GMD-1B) hosted in gas filled balloons, (iv) thermoscreen (v) pilot balloons to for cloud ceiling measurements. Equipped with corner reflectors these balloons are tracked by theodolites, optical trackers and radars.

The meteorological data are processed together with information from other forecasting services and output to the Range Officer is provided by the Range Meteorological Staff Office.

In particular wind and thermal constraints must be met: Fig. 1.53 shows the limits of wind intensity and direction in relation to the possible bending effect on the launcher structure. Concentric circles define winds of equal speed from all the directions. The launcher azimuth is 90 deg (eastward launch), so the headwind is 90 deg and the tailwind is 270 deg. The range of allowable wind speed lies within the oval curves associated with different altitudes: up to 8 km (solid line), 10.5 km (dashed line) and 14 km (dotted line). The Figure shows the wind speed curves for 100 knots = 51 m/s, 200 knots = 101 m/s and 300 knots = 153 m/s.

A second wind constraint is due to the aerodynamic control of the launcher during the first stage flight. Wing tips will exceed the maximum deflection if the wind intensity and direction are outside the closed curve of Fig. 1.54 related to altitudes of 3 to 4 km.

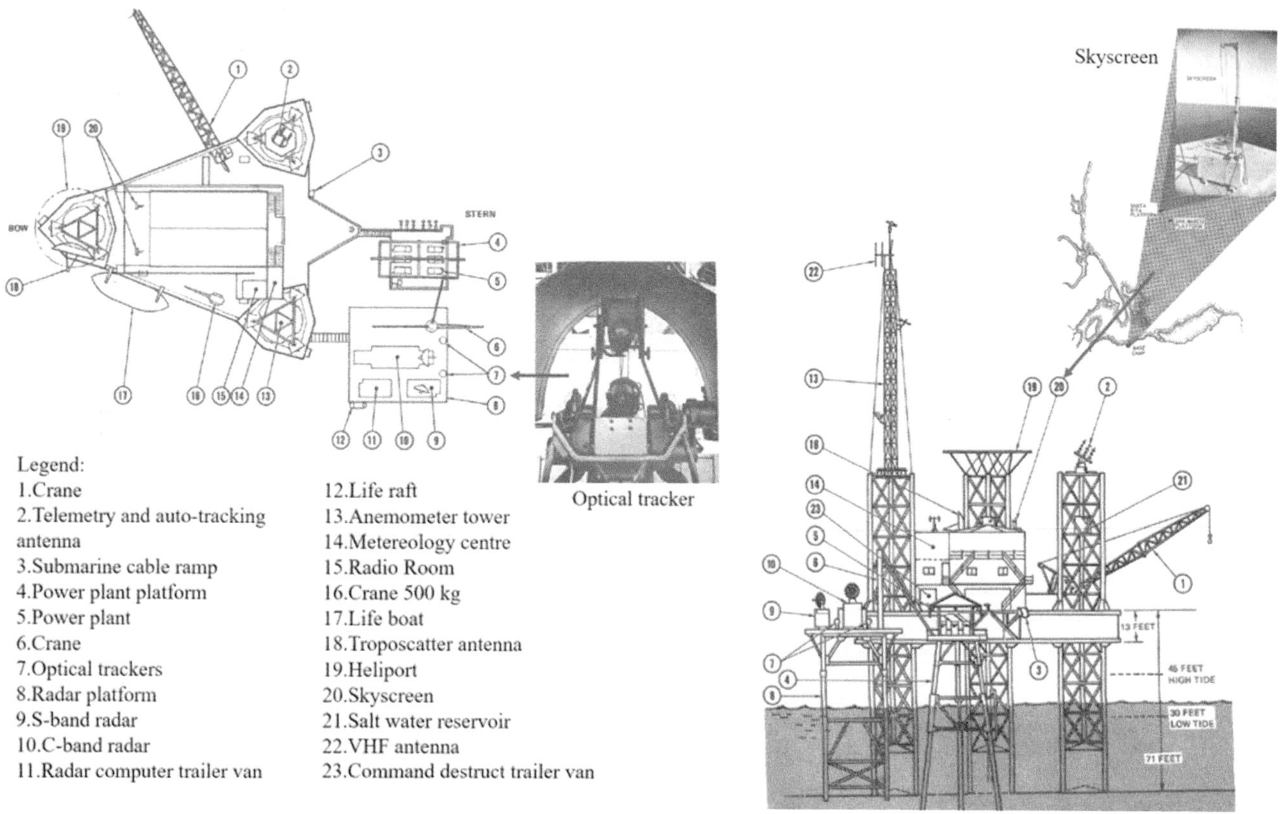

Fig. 1.52 Assembly of the Santa Rita platform

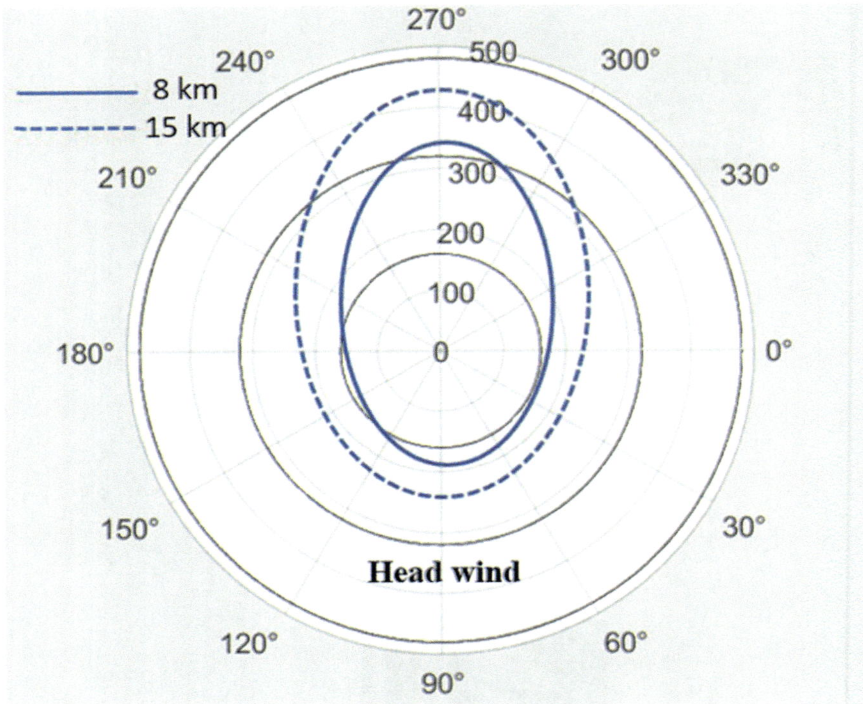

Fig. 1.53 Allowed regions (inside the closed curves) of wind velocity and directions at different altitudes: bending moment constraint [19]

Thermal constraints must also be considered by the Range Safety Officer in order to proceed with launch operations. The conditions are given according to the thermal exposure of the rocket at different temperatures for different durations. In Fig. 1.55 the countdown of the launcher can continue without interruption if the exposure of the launcher is in Area I. If the rocket exposure is in Area II a reconditioning for 120 h at controlled temperature between 21 and 32 deg is required before firing, if in Area III the mission must be stopped.

The Santa Rita platform houses all the equipment for remote vehicle checkout and launch.

Telemetry from the launcher is received by the auto tracking VHF antenna with control room located in a trailer van on in the main deck of the Santa Rita platform. Remote commands for the final tests and operations are given from the blockhouse, see Fig. 1.56 for the vehicle control room assembly.

The initial phases of the lift-off trajectory are monitored by optical facilities such as the optical tracker and the skyscreens. Soon after the vertical arc, the trajectory is monitored by two radar antennas, one S-band and the other C-band.

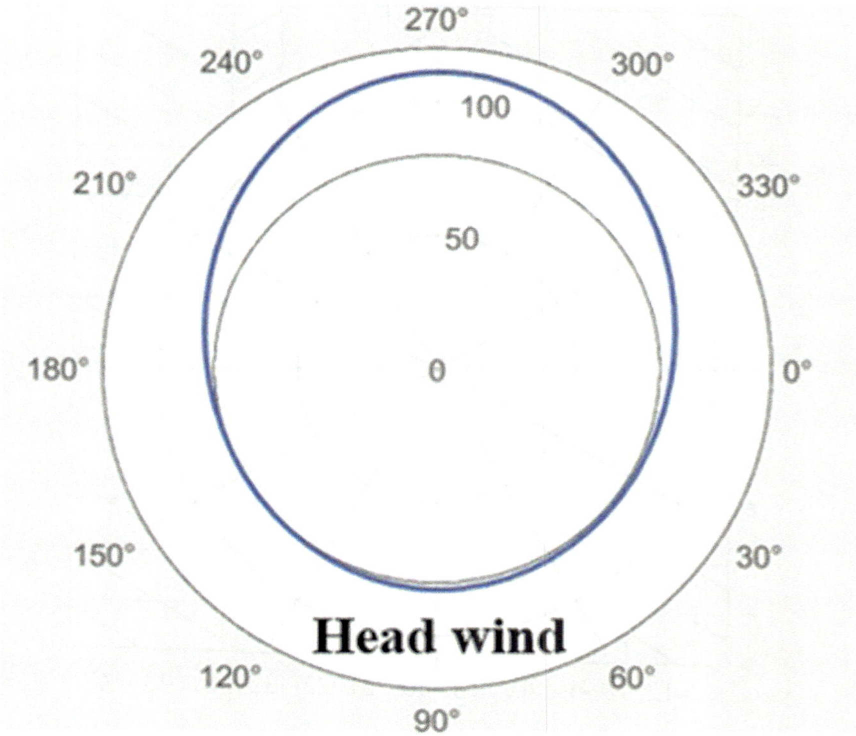

Fig. 1.54 Allowed regions (inside the closed curves) of wind velocity and directions at altitudes within 3 to 4 km: control angles constraint [19]

The radar data are processed and displayed on two plotboards for each radar signal. One plotboard shows the vertical projection of the trajectory (Z vs. time), the second shows the horizontal projection (X vs. Y).

Telemetry flight data are also collected at the blockhouse control centre by further 3 plotboards and and an 8-channel strip chart recorder. Telemetry data are: antenna elevation and azimuth look angles versus time, longitudinal acceleration, pitch and yaw angles versus time. These data are analysed in real time by the Range Safety Officer to assess whether a destruct command needs to be generated. This command is transmitted to the rocket by a high reliability transmitter based on a UHF antenna and it is used to activate the vehicle's destruct charges.

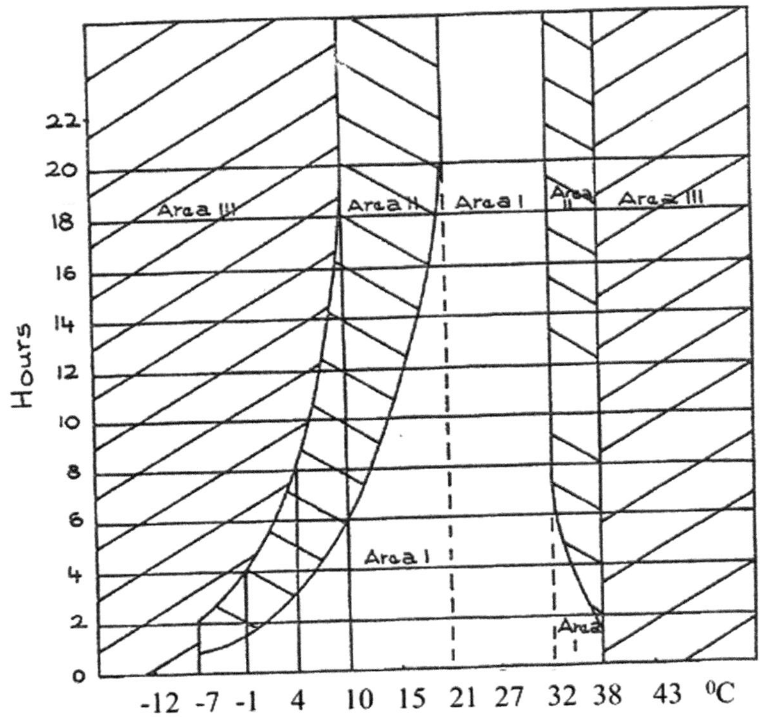

Fig. 1.55 Time limits for launcher exposure temperature [19]

Problems

1.1 Determine the azimuth angle to reach an orbit with inclination i = 40 deg from the San Marco Platform.

1.2 The mission of the Prob.1.1 has launch date T_0 = 26 October 2026: determine the RAAN angle Ω achieved at launch time $t^* = 6\,h, 0\,min, 0\,s$.

Solutions

Problems of This Chapter

1.1 San Marco Platform has longitude and latitude equal to: $\lambda_g = 40.2\,\text{deg}\,E$, $L_0 = 2.93\,\text{deg}\,S$. For i = 40 deg the Launch Base Eq. (1.26) gives: $\psi_A = asin(\frac{cos(i)}{cos L_0})$. The two solutions are $\psi_A = 50°$ and $\psi_A = 130°$, see Fig. 1.32.

If the latter solution is considered (descending node), the correction needed to compensate the Earth eastward velocity gives $\psi_A = 132.7°$, see Problem 2.6.

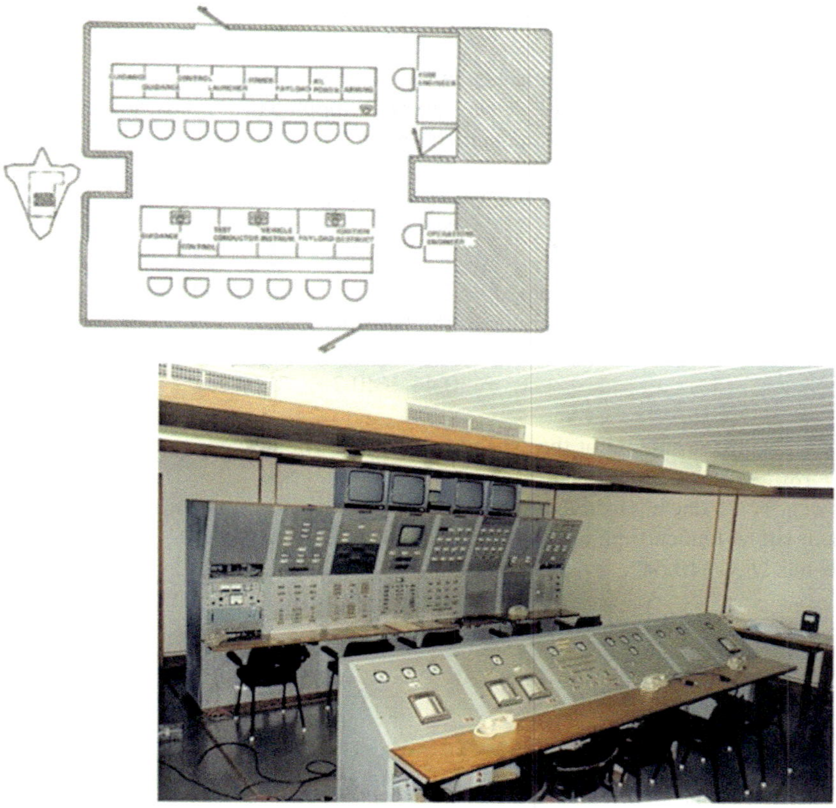

Fig. 1.56 Assembly of the vehicle control center

1.2 The RAAN is found according to Eq. (1.30). The absolute Greenwich longitude λ_{G0} at date T_0 is calculated.

Let y be the year, m be the month, d be the day corresponding to T_0. Its Julian date is computed by the steps:

- $a_1 = d - 32075$
- $a_2 = fix((m - 14)/12)$ where fix(x) rounds x to the nearest integer towards zero
- $a_3 = 1461 * (y + 4800 + a_2)$
- $a_4 = 367 * (m - 2 - a_2 * 12)$
- $a_5 = 3 * fix((y + 4900 + a_2)/100)$
- $a_6 = a_1 + fix(a_3/4) + fix(a_4/12) - fix(a_5/4)$

The Julian date is $J = a_6 - 0.5$, $\tau = (J - 2415020.0)/36525$ and the absolute Greenwhich longitude at T_0 is:

$$\lambda_{G0} = 99.6909832 + 36000.7689 * \tau + 0.00038708 * \tau^2$$

modulo 360 deg, so that the angle λ_{G0} in the range [0°, 360°].

The time of the date $t_0 = hr + min + sec$ is computed in seconds, so the absolute longitude of the launch base at the launch time is:

$$\lambda_a = \lambda_{G0} + \omega_E\, t + \lambda_g$$

mod-360, when necessary.

For the date T_0 of the problem, $J = 2.4613\ 10^6$, so $\lambda_{G0} = 30.44$ deg.

– at $t_0 = 0$ the absolute longitude of the launch site is:

$$\lambda_a = \lambda_{G0} + \lambda_g = 30.44 + 40.2 = 70.64 \text{ deg}$$

This is also the argument of the ascending node Ω. The descending node has angle: $\Omega = 180° + 70.64° = 250.64°$.

– at $t_0 = 6\,h$ one has to add the angle $\omega_E\,\tau = 90°$, and the absolute longitude of the launch site at launch time is $\lambda_a = 160.64$ deg, equal to the RAAN. The descending node has $\Omega = 340.64°$.

References

1. Scout Planning guide, LTV Missile and Electronics Division (1987)
2. Tanck, P., Williams, J.: The Scout Launch vehicle system (1988). https://digitalcommons.usu.edu/cgi/viewcontent.cgi?article=2401context=smallsat
3. Scout performances and trajectory report, Vough Corporation Report n.23.395 (1977)
4. The Scout G1 launch vehicle, LTV Aerospace and Defence (1987)
5. Leiss, A.: Scout Vehicle program, Phase 6 1982, republished as nasa/techdoc/19820073028 in https://archive.org/ (2010)
6. Prussing, J., Conway, B.: Orbital Mechanics. Oxford University Press (1993)
7. Teofilatto, P., Carletta, S., Pontani, M: Analytic derivation of ascent trajectories and performance of launch vehicles. Appl. Sci. **12**(11) (2022)
8. Vega User's manual, Issue 4, Ariane Space (2014). http://www.arianespace.com
9. Avio: our products Vega E, https://www.avio.com/it/vega-e (2022)
10. Scout rocket motor and pyrotechnics parameter summary, Vough Corporation (1977)
11. Scout training manual, vol. 1 Mechanical Systems, ELV Report (1962)
12. Ferrer Jara, G.: Safe and Arm systems for small space launchers. Thesis for the Special Master in Aerospace Engineering, School of Aerospace Engineering of Rome (2019)
13. Scout launch vehicle illustrated components and schematic dyagrams, ELV Report (1984)
14. Scout System Design Report, R-ED 11122, Honeywell (1963)
15. Blau, P.: Pegasus XL launch vehicle. Space news and beyond, spaceflight, Launch vehicle library (2017)
16. China's commercial rocket launches satellites from sea, Huaxia. https://english.news.cn/20240529/f00d7d839986460f93bc589098a71928/c.html (2024)
17. Broglio, L.: The future of the San Marco Project. In: Proceedings of the XXVIII COSPAR Conference (1990)
18. Sirinian, M.: La traiettoria di lancio del San Marco II, Atti del CRA, vol. 15g (1967)

19. Scout Operation plan, CRA Report (1971)
20. Scout San Marco Standard Operating Procedures, vol. III, book 2, CRA (1984)

Chapter 2
Ascent Trajectories: Atmospheric Arc

Abstract The atmospheric arc of a space trajectory has a strong path constraint due to the aerodynamic lateral loads on the structure. This forces the trajectory to follow a path with a zero angle of incidence (gravity turn trajectory). This chapter introduces analytical formulae for the gravity turn of the launcher. These formulas are useful to evaluate the variation of velocity loss due to gravity and to provide an estimate of the launcher performance. A simple generalisation of the well-known Tziolkowski formula for the rapid evaluation of launcher capability is introduced, which allows a more accurate estimate of the payload mass that a launcher can place in orbit. The analytical estimate is compared with the highly accurate numerical model of the ascent trajectory in the atmosphere, which is derived in detail. The different phases of the launcher trajectory are analysed: vertical ascent, pitch manoeuvre, pitch over, gravity turn. The general equations of motion of a launcher in the atmosphere are derived. The guidance parameters of the trajectory in the atmosphere are introduced and it is shown that the main parameter influencing the atmospheric arc is the duration of the pitch manoeuvre that the launcher has to perform after the vertical ascent.

2.1 Introduction

In principle there are many trajectories that a launcher can follow to put a satellite into orbit. These trajectories differ in the direction of the thrust, in the thrust intensity, in the number and the duration of the coasting arcs, in a word, in the launcher's guidance law [1, 2]. The guidance law must be such that the final state of the payload (the spacecraft) coincides with the target orbital state while satisfying any constraints that are imposed during the flight, such as:

- Launch site safety conditions
- Re-entry of the exhausted stages in the prescribed areas
- Avoiding flying over prohibited areas (populated areas, areas belonging to different countries, etc.)
- Limiting the maximum aerodynamic load
- Limit on the maximum mechanical load

P. Teofilatto, *Trajectories of Launch and Reentry Systems*, Springer Aerospace Technology, https://doi.org/10.1007/978-3-031-75760-0_2

- Limit on the maximum thermal flux during atmospheric ascent
- Trajectory constraints due to the visibility from ground stations.

From mathematical point of view the design of a launcher trajectory can be formulated as an optimal control problem [3]:

Starting from the initial launch conditions $\mathbf{x}(t_0) = \mathbf{x}_0$, *determine the guidance law* $\mathbf{u}(t)$, *that is the time variation of the thrust direction and intensity, in order to achieve prescribed final conditions (injection into orbit)* $x_{1_f} \ldots x_{k_f}$, *while satisfying some flight constraints* $\boldsymbol{\psi}(\mathbf{x}, \mathbf{u}, t) = 0$ *and minimizing a cost function*

$$J = \phi(\mathbf{x}_f, t_f) + \int_0^{t_f} L(\mathbf{x}(t), \mathbf{u}(t), t)\, dt$$

depending on the final state $\mathbf{x}_f$, *the final time* t_f *and a function L depending on the state variables* $\mathbf{x}$ *and controls* $\mathbf{u}$ *defined along the flight.*

In fact in the atmospheric arc the launch trajectory is highly constrained because the aerodynamic loads on the launcher structure force the guidance system to follow a zero angle of incidence ($\alpha = 0$). This is related to the flexibility of the launcher: to reduce the mass of the structure as much as possible, in fact the launcher is a slender flexible body capable of withstanding high axial loads (generated by the thrust and the aerodynamic drag), but allowing low lateral loads. Lateral loads are generated by the lift which is proportional to the product $q\alpha$ between the dynamic pressure $q = \frac{1}{2}\rho\, V_R^2$ and the angle of incidence α. Namely there is a "*q times alpha*" constraint $q\alpha \le q\alpha_{max}$ that must be satisfied. During the flight the dynamic pressure q is limited in two circumstances: in the initial phase of the launch, when the velocity V_R is small and in the exo-atmospheric flight, when the rocket reaches high altitudes and the density ρ becomes exponentially small. Between these two conditions, the dynamic pressure q is quite high and the rocket must follow a path with an angle of attack α close to zero to satisfy the constraint. When α is set to zero, this arc of the trajectory is called gravity turn.

In fact, it is possible to introduce the $q\,\alpha$ constraint in the setting of the optimal control problem. The constrained problem is rather difficult to solve, although progress has been made by algorithms based on: (i) the indirect approach, through the determination of the lagrangian multipliers introduced to cope with the constrained optimal problem [4–7], (ii) the direct approach, through the discretisation of the problem see [8, 9], (iii) indirect heuristic approach, through a smart search of the initial conditions of the lagrangian multipliers, [10, 11].

The numerical results generally show that there are small improvements in the cost function by pursuing an optimal path satisfying the atmospheric load constraint with respect to the gravity turn arc, so the latter will be considered in the following and the output will be a feasible sub-optimal trajectory divided into the phases:
(i) vertical ascent
(ii) pitch manoeuvre
(iii) gravity turn.

This Chapter defines the equation of motion and the guidance law of these phases, introducing different reference frames and the most appropriate state variables for each phase. Some analytical solutions for the gravity turn arc are also presented which are useful to evaluate in very short time the trajectory of a space launcher with a given flight path angle γ_{ex} at the exit of the atmospheric arc. It is shown that the atmospheric trajectories essentially depend on the single guidance parameter γ_{ex} which corresponds to the duration t_p of the pitch manoeuvre. Different values of t_p produce different conditions at the end of the atmospheric ascent, and hence different end conditions of the launcher mission. The analytical formulae introduced here allow the choice of the correct guidance parameter t_p for injection into the target orbit.

2.2 Lift-Off and Vertical Ascent

In general a launcher follows a vertical trajectory at lift-off. From an energetic point of view the vertical trajectory is the less efficient since the gravitational acceleration is completely opposite to the variation of the velocity. However a vertical trajectory is necessary at the beginning of the launch for safety reasons. In fact the large flow of exhaust gases must be contained and possibly addressed within channels inside the ground, and a vertical launch avoids the dispersion of the gases on the launch pad.

For instance the San Marco platform had an opening through the platform hull located beneath the launcher to allow a free discharge path to the sea of the vehicle exhaust gasses.

Another important point is the "gravity turn" effect: if the trajectory is not vertical the effect of the gravity is "to turn" the direction of the speed downwards, as it can be seen from a physical point of view simply by observing the "parabola" that a body follows under the effect of gravity alone. This effect is expressed by the formula:

$$\dot{\gamma}_R = -\frac{g \sin \gamma_R}{V_R} \tag{2.1}$$

where γ_R is the flight path angle, that is the angle between the direction of the launcher's relative velocity V_R and the horizontal direction and g is the acceleration due to gravity. In the vertical arc $\gamma_R = \frac{\pi}{2}$ then $\dot{\gamma} = 0$ and the trajectory remains vertical. At lift-off the velocity V_R is low, so the rate of change of the velocity direction $\dot{\gamma}_R$ is high and there is a risk of turning down and hitting the ground in a short time. On the other hand, when the launcher reaches a significant velocity in vertical flight, the term (2.1) becomes smaller and the launcher can deviate from its vertical trajectory without danger.

The duration t_v of the vertical arc is as short as possible and it is determined by the launcher's ability to increase its speed (thrust to weight ratio n) and by the safety requirements of the ground station.

The equations of the vertical arc are described here in the Launch Base Inertial frame (LBI) defined in Sect. 1.9 by the axes ($\hat{r}_0$, $\hat{E}_0$, $\hat{N}_0$), where $\hat{r}_0$ is the radial direction and $\hat{E}_0$, $\hat{N}_0$ are the local East and the local North directions with respect to the launch base position at the time t_0 of the lift off.

Then relationship between the Earth Centered Inertial (ECI) to LBI axes is given by the two rotations of absolute longitude λ_{a_0} around the $\hat{c}_3$ axis of ECI frame and latitude L_0 around the new axis 2 in clockwise (negative) direction. In matrix form:

$$\begin{bmatrix} \hat{r}_0 \\ \hat{E}_0 \\ \hat{N}_0 \end{bmatrix} = R^2_{-L_0} R^3_{\lambda_{a0}} \begin{bmatrix} \hat{c}_1 \\ \hat{c}_2 \\ \hat{c}_3 \end{bmatrix} \tag{2.2}$$

The kinematic and dynamic equations

$$\frac{d\mathbf{r}}{dt} = \mathbf{V}$$

$$\frac{d\mathbf{V}}{dt} = \frac{\mathbf{F}}{m}$$

are written with respect to the "relative to the wind velocity" $\mathbf{V}_R$ This is defined by:

$$\mathbf{V}_R = \mathbf{V} - [\boldsymbol{\omega}_E \wedge \mathbf{r} + \mathbf{w}_{loc}]$$

where $\mathbf{V}$ is the absolute velocity and the velocity in square brackets is the wind velocity considered as the sum of the velocity of the atmosphere rotating with the Earth angular velocity $\boldsymbol{\omega}_E$ plus the velocity of the local winds $\mathbf{w}_{loc}$ flowing on the launch site (the velocity $\mathbf{w}_{loc}$ will be neglected in the following derivations).

The relative velocity $\mathbf{V}_R$ is preferred to the absolute velocity $\mathbf{V}$ because the aerodynamic forces and the $q\alpha$ constraint depend on the relative velocity.

Then the kinetic and dynamic equations in the relative velocity are:

$$\frac{d\mathbf{r}}{dt} = \mathbf{V} = \mathbf{V}_R + \boldsymbol{\omega}_E \wedge \mathbf{r} \tag{2.3}$$

$$\frac{d\mathbf{V}}{dt} = \frac{d}{dt}(\mathbf{V}_R + \boldsymbol{\omega}_E \wedge \mathbf{r}) = \frac{\mathbf{F}}{m}$$

Recall that $\boldsymbol{\omega}_E = \omega_E\, \hat{c}_3$ and ω_E and $\hat{c}_3$ are constant, so the last equation is equivalent to:

$$\frac{d\mathbf{V}}{dt} = \frac{d\mathbf{V}_R}{dt} + \boldsymbol{\omega}_E \wedge \mathbf{V} = \frac{\mathbf{F}}{m}$$

that is:

$$\frac{d\mathbf{V}}{dt} = \frac{d\mathbf{V}_R}{dt} + \boldsymbol{\omega}_E \wedge (\mathbf{V}_R + \boldsymbol{\omega}_E \wedge \mathbf{r}) = \frac{\mathbf{F}}{m}$$

and the dynamic equation is:

$$\frac{d\mathbf{V}_R}{dt} + \boldsymbol{\omega}_E \wedge \mathbf{V}_R + \boldsymbol{\omega}_E \wedge (\boldsymbol{\omega}_E \wedge \mathbf{r}) = \frac{\mathbf{F}}{m} \tag{2.4}$$

Let the position and velocity be in LBI frames:

$$\mathbf{r} = (X_1, X_2, X_3)\,, \;\; \mathbf{V}_R = (V_1, V_2, V_3)$$

In (2.3), the only vector which is not in LBI coordinates is $\boldsymbol{\omega}_E = \omega_E \hat{c}_3$.

By inversion of (2.2) we have

$$\hat{c}_3 = \sin L_0 \hat{r}_0 + \cos L_0 \hat{N}_0 \tag{2.5}$$

then

$$\boldsymbol{\omega}_E \wedge \mathbf{r} = \omega_E \begin{bmatrix} \hat{r}_0 & \hat{E}_0 & \hat{N}_0 \\ \sin L_0 & 0 & \cos L_0 \\ X1 & X2 & X_3 \end{bmatrix}$$

i.e.

$$\boldsymbol{\omega}_E \wedge \mathbf{r} = -\omega_E\, X_2 \cos L_0 \hat{r}_0 + \omega_E\, (X_1 \cos L_0 - X_3 \sin L_0) \hat{E}_0 + \omega_E\, X_2 \sin L_0 \hat{N}_0 \tag{2.6}$$

Then the kinetic equations (2.3) give the scalar equations

$$\begin{aligned} \tfrac{dX_1}{dt} &= V_1 - \omega_E\, X_2 \cos L_0 \\ \tfrac{dX_2}{dt} &= V_2 + \omega_E\, (X_1 \cos L_0 - X_3 \sin L_0) \\ \tfrac{dX_3}{dt} &= V_3 + \omega_E\, X_2 \sin L_0 \end{aligned} \tag{2.7}$$

To derive the scalar equations of (2.4), note that the term $\boldsymbol{\omega}_E \wedge \mathbf{V}_R$ has the same expression of (2.6) with V_1, V_2, V_3 instead of X_1, X_2, X_3.

By triple vector formula the other term in the left hand side of (2.4) is:

$$\boldsymbol{\omega}_E \wedge (\boldsymbol{\omega}_E \wedge \mathbf{r}) = (\boldsymbol{\omega}_E \mathbf{r}) \boldsymbol{\omega}_E - \omega_E^2\, \mathbf{r}$$

with

$$(\boldsymbol{\omega}_E \mathbf{r})\, \boldsymbol{\omega}_E = \omega_E^2 (\sin L_0 X_1 + \cos L_0 X_3)(\sin L_0 \hat{r}_0 + \cos L_0 \hat{N}_0)$$

and

$$\omega_E^2\, \mathbf{r} = \omega_E^2 X_1 \hat{r}_0 + \omega_E^2 X_2 \hat{E}_0 + \omega_E^2 X_3 \hat{N}_0$$

Then the scalar equations of the dynamics are:

$$\begin{aligned}
&\tfrac{dV_1}{dt} - V_2\omega_E \cos L_0 + r\omega_E^2(\sin L_0 X_1 + \cos L_0 X3)\sin L_0 - \omega_E^2 X_1 = \tfrac{F_1}{m} \\
&\tfrac{dV_2}{dt} + V_1\omega_E \cos L_0 - V_3\omega_E \sin L_0 - \omega_E^2 X_2 = \tfrac{F_2}{m} \\
&\tfrac{dV_3}{dt} + V_2\omega_E \sin L_0 + r\omega_E^2(\sin L_0 X_1 + \cos L_0 X3)\cos L_0 - \omega_E^2 X_3 = \tfrac{F_3}{m}
\end{aligned} \tag{2.8}$$

The initial conditions are (l distance of the launcher centre of mass at lift off):

$$\mathbf{r}_0 = (R_E + l, 0, 0)\ ,\quad \mathbf{V}_{R_0} = (0, 0, 0) \tag{2.9}$$

and the duration of the vertical trajectory is t_v.

In (2.8) the accelerations ($\frac{F_1}{m}, \frac{F_2}{m}, \frac{F_3}{m}$) are the three components of gravity, aerodynamic and thrust accelerations:

$$\mathbf{g} + \frac{D}{m}(-\hat{v}) + \frac{L}{m}\hat{l}_a + \frac{T}{m}\hat{T}$$

The gravity acceleration has the simple expression:

$$\mathbf{g} = -\frac{\mu}{r^3}\mathbf{r} = -\frac{\mu}{r^3}(X_1\hat{r}_0 + X_2\hat{E}_0 + X_3\hat{N}_0) \tag{2.10}$$

The drag acceleration is:

$$\frac{\mathbf{D}}{m} = -\frac{1}{2}\rho V_R^2 \frac{C_D S}{m}\,\hat{v} \tag{2.11}$$

with

$$\hat{v} = \frac{V_1}{V_R}\hat{r}_0 + \frac{V_2}{V_R}\hat{E}_0 + \frac{V_3}{V_R}\hat{N}_0 \tag{2.12}$$

The lift acceleration is:

$$\frac{\mathbf{L}}{m} = \frac{1}{2}\rho V_R^2 \frac{C_L S}{m}\,\hat{l}_a \tag{2.13}$$

and $\hat{l}_a$ is defined later in (2.16).

Since the body is axisymmetric, the drag and the lift coefficients C_D, C_L depend on the "total" angle of attack α between the longitudinal axis of the launcher $\hat{x}_b$ and the relative velocity direction $\hat{v}$, see Fig. 2.1. In the same Figure, the angles β (sideslip angle) and α_π (incidence angle) are shown defining the relative velocity direction in the planes ($\hat{y}_B, \hat{z}_B$), ($\hat{x}_B, \hat{y}_B$) respectively.

Guidance is given by the thrust direction $\hat{T}$ which is along the vertical direction. Sometimes the launch tower is tilted by a kick angle χ_0 and rotated by the azimuth angle ψ_A. In this case the thrust direction is equal to:

$$\hat{T} = \cos\chi_0\hat{r}_0 + \sin\chi_0\sin\psi_A\hat{E}_0 + \sin\chi_0\cos\psi_A\hat{N}_0 \tag{2.14}$$

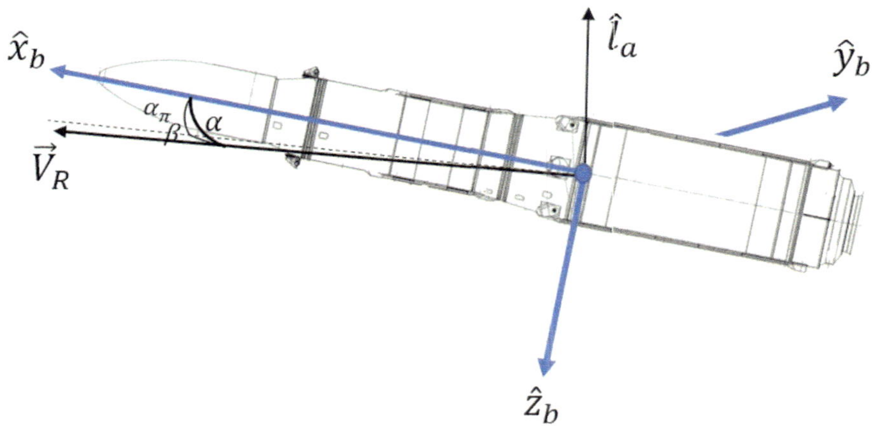

Fig. 2.1 The rocket body axes $\hat{x}_b$, $\hat{y}_b$, $\hat{z}_b$, the lift direction $\hat{l}_a$ and the total angle of attack α

Note that $\hat{x}_b = \hat{T}$ and both $\hat{x}_b$ and $\hat{v}$ are already defined in the LBI frame, then it is easy to obtain the scalar product

$$\cos\alpha = \hat{x}_b\,\hat{v}$$

to get the angle of attack α:

$$\alpha = acos(\hat{x}_b\,\hat{v}) \tag{2.15}$$

Consider now the vector $\mathbf{h}_a$ orthogonal to the plane defined by $\hat{x}_b\hat{v}$: $\mathbf{h}_a = \hat{x}_b \wedge \hat{v}$.

The lift is aligned to the vector

$$\mathbf{l}_a = \mathbf{h}_a \wedge \hat{v} = -\hat{v} \wedge (\hat{x}_b \wedge \hat{v}) = -\hat{x}_b + (\hat{v}\dot{\hat{x}}_b)\hat{v} = \cos\alpha\hat{v} - \hat{x}_b$$

whose versor

$$\hat{l}_a = \frac{\cos\alpha}{\sin\alpha}\hat{v} - \frac{1}{\sin\alpha}\hat{x}_b \tag{2.16}$$

defines the lift direction.

Recall that the drag coefficient was defined in Sect. 1.7 by the "polar equation"

$$C_D = C_{D_0} + k_D C_L^2$$

where the coefficient C_{D_0} is the drag coefficient at zero angle of attack and the lift coefficient C_L is regarded as a linear function of α:

$$C_L = C_{L_\alpha}\alpha$$

The aerodynamic coefficients C_{D_0}, k_D and C_{L_α} depend on Mach number and interpolation formulae relative to the SCOUT launch vehicle are given in the solution to Problem 2.1.

The thrust direction (2.14) was defined by the kick and azimuth angles. In general however this direction (that is the direction of $\hat{x}_b$) is defined by the yaw, pitch and roll attitude angles of a launcher: ψ, θ and ϕ.

In order to be consistent with the conventional yaw-pitch-roll rotations, the LBI frame is transformed into an auxiliary reference frame AUX obtained by a permutation of indices:

$$\begin{bmatrix} \hat{i}_0 \\ \hat{j}_0 \\ \hat{k}_0 \end{bmatrix} = \begin{bmatrix} 0 & 1 & 0 \\ 0 & 0 & -1 \\ -1 & 0 & 0 \end{bmatrix} \begin{bmatrix} \hat{r}_0 \\ \hat{E}_0 \\ \hat{N}_0 \end{bmatrix} \tag{2.17}$$

Then the rocket body axes $(\hat{x}_b, \hat{y}_b, \hat{z}_b)$ are obtained by the 3-2-1 rotations:

$$\begin{bmatrix} \hat{x}_b \\ \hat{y}_b \\ \hat{z}_b \end{bmatrix} = R^1_\phi \, R^2_\theta \, R^3_\psi \begin{bmatrix} \hat{i}_0 \\ \hat{j}_0 \\ \hat{k}_0 \end{bmatrix} \tag{2.18}$$

The **Body Frame** (BF) $(\hat{x}_b, \hat{y}_b, \hat{z}_b)$ is a reference frame with the origin at the rocket centre of mass, $\hat{x}_b$ corresponds to the rocket longitudinal axis and $\hat{y}_b$, $\hat{z}_b$ are on the rocket symmetry plane.

In vertical flight one has $\hat{T} = \hat{x}_b$: a relationship between the attitude angles and the kick and azimuth angle in (2.14) is derived.

According to Sect. 1.9 the azimuth angle is determined by the launch site latitude and inclination of the target orbit. The selection of the azimuth angle ψ_A is the first step in determining the thrust direction. The azimuth angle corresponds to a first rotation around $\hat{k}_0$ of the yaw angle $\psi_0 = \psi_A - \frac{\pi}{2}$.

After such a rotation, the axis 2 becomes the pitch axis $\hat{y}_b$ and the initial pitch angle θ_0 around $\hat{y}_b$ determines the thrust direction. If the thrust direction is vertical $\theta_0 = \frac{\pi}{2}$, if there is a kick angle χ_0, then $\theta_0 = \frac{\pi}{2} - \chi_0$.

It is now shown that the thrust direction defined by the previous yaw and pitch angles corresponds to the direction (2.14) defined by the azimuth and kick angles.

$$\begin{bmatrix} \hat{x}_b \\ \hat{y}_b \\ \hat{z}_b \end{bmatrix} = R^2_{\theta_0} \, R^3_{\psi_0} \begin{bmatrix} \hat{i}_0 \\ \hat{j}_0 \\ \hat{k}_0 \end{bmatrix} = \begin{bmatrix} c_{\theta_0} & 0 & -s_{\theta_0} \\ 0 & 1 & 0 \\ s_{\theta_0} & 0 & c_{\theta_0} \end{bmatrix} \begin{bmatrix} c_{\psi_0} & s_{\psi_0} & 0 \\ s_{\theta_0} & c_{\psi_0} & 0 \\ 0 & 0 & 1 \end{bmatrix} \begin{bmatrix} \hat{i}_0 \\ \hat{j}_0 \\ \hat{k}_0 \end{bmatrix}$$

That is

$$\begin{bmatrix} \hat{x}_b \\ \hat{y}_b \\ \hat{z}_b \end{bmatrix} = \begin{bmatrix} c_{\theta_0} c_{\psi_0} & c_{\theta_0} s_{\psi_0} & -s_{\theta_0} \\ -s_{\psi_0} & c_{\psi_0} & 0 \\ s_{\theta_0} c_{\psi_0} & s_{\theta_0} s_{\psi_0} & c_{\theta_0} \end{bmatrix} \begin{bmatrix} 0 & 1 & 0 \\ 0 & 0 & -1 \\ -1 & 0 & 0 \end{bmatrix} \begin{bmatrix} \hat{r}_0 \\ \hat{E}_0 \\ \hat{N}_0 \end{bmatrix}$$

Then

$$\hat{T} = \hat{x}_b = \sin\theta_0 \hat{r}_0 + \cos\theta_0 \cos\psi_0 \hat{E}_0 - \cos\theta_0 \cos\psi_0 \hat{N}_0 \tag{2.19}$$

Considering that

$$\theta_0 = \frac{\pi}{2} - \chi_0 \;\; and \;\; \psi_0 = \psi_A - \frac{\pi}{2}$$

we get that the formula (2.19) is equal to the formula of the thrust direction (2.14).

2.3 Pitch Manoeuvre

At time t_v the vertical arc ends and the launcher must rotate to the orbit plane selected by the azimuth angle ψ_A defined in Sect. 1.9. Such a rotation is, by convention, a rotation around the launcher pitch axis $\hat{y}_b$.

During the vertical launch the launcher's longitudinal axis, the thrust and the relative velocity are all aligned and there is no angle of attack α. When the launcher is rotated in pitch by an attitude command, there is a misalignment between the longitudinal axis and the direction of the velocity, hence an angle $\alpha \neq 0$ is generated. The pitch manoeuvre ends (pitch over) when such incidence angle returns to zero. This happens for two effects: thrust and gravity. Namely, after the rocket pitch rotation the thrust is oriented along a new direction $\hat{x}_b$. At time steps $t_k = t_{k-1} + \Delta t$ the current velocity $\mathbf{V}_k$ is the sum of the velocity $\mathbf{V}_{k-1}$ and the velocity $\Delta\mathbf{V} = \frac{T}{m}\,\Delta t\,\hat{x}_b$ generated by the thrust. Step by step the resultant direction continuously approaches the new thrust direction $\hat{x}_b$ and the angle α tends to zero, see Fig. 2.2.

Moreover, as soon as the launcher deviates from the vertical direction, the gravity will turn the trajectory away from the vertical direction and the velocity, "turned by the gravity", will reach the launcher's axis $\hat{x}_b$, see Fig. 2.3.

When the speed direction becomes parallel to $\hat{x}_b$, $\hat{v} = \hat{x}_b$, the pitch over occurs and the angle between the velocity of the launcher and the vertical direction is called **kick angle**. This is a crucial parameter in the launch trajectory. The value of the kick angle is determined by the execution of the pitch manoeuvre.

The behaviour of the α angle during the pitch manoeuvre is shown in Fig. 2.4. After the vertical trajectory at zero α angle, the manoeuvre generates a (negative) angle α. Although significant α angles can be reached during the manoeuvre, the $q\,\alpha$ constraint is not violated due to the reduced value of the velocity and hence of the dynamic pressure q.

The pitch manoeuvre is a commanded pitch angle $\theta(t)$, $t > t_v$ following two laws:

1. For $t \in [t_v, t_v + t_p]$ the pitch is decreased linearly:

$$\theta(t) = \theta_0 - \dot{\theta}_0(t - t_v) \tag{2.20}$$

2. For $t \in [t_v + t_p, t_v + t_p + t_{po}]$ the pitch angle is kept constant

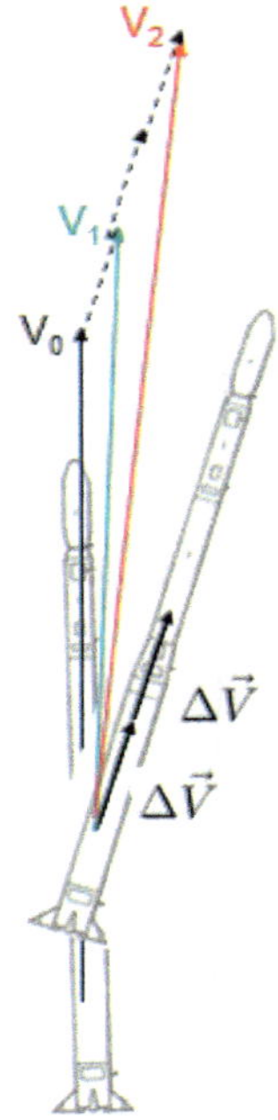

Fig. 2.2 The reduction of α by the thrust effect

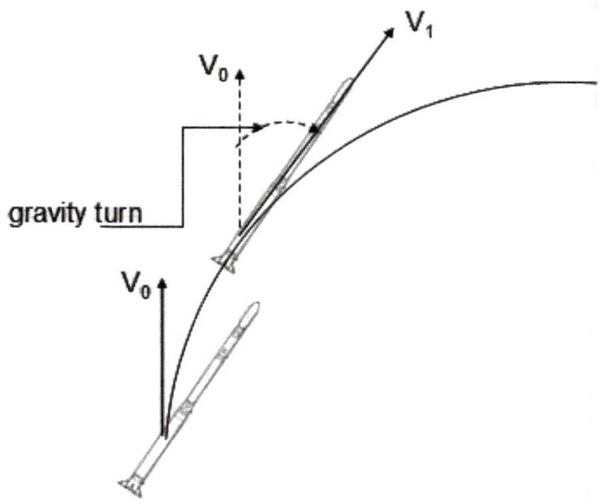

Fig. 2.3 The reduction of α by the gravity effect

$$\theta(t) = \theta_0 - \dot{\theta}_0 t_p \tag{2.21}$$

The equations of motion are again (2.7), (2.8) with thrust direction $\hat{T}$

$$\hat{T} = \sin\theta \hat{r}_0 + \cos\theta \cos\psi \hat{E}_0 - \cos\theta \cos\psi \hat{N}_0 \tag{2.22}$$

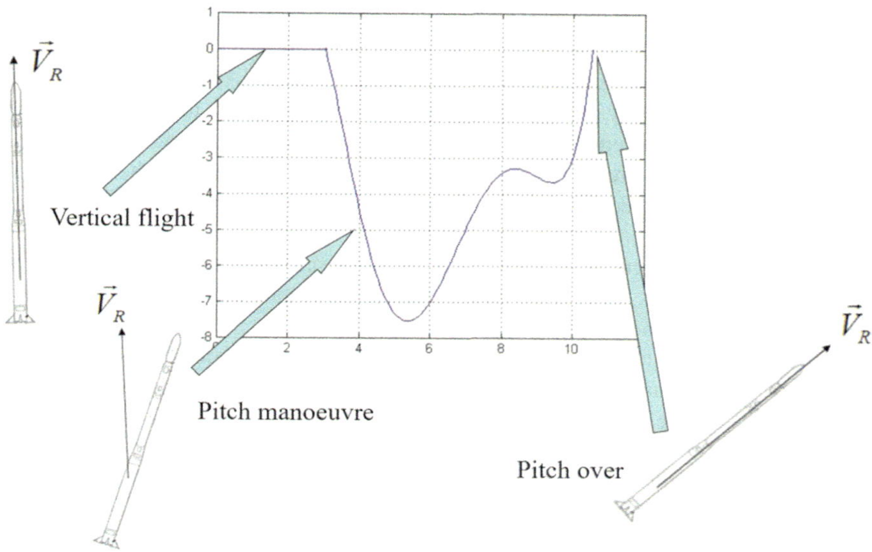

Fig. 2.4 The behaviour of α in time

with θ equal to the laws 1. and 2. (see formulas (2.20), (2.21)) and $\psi = \psi_A - \frac{\pi}{2}$.

Manoeuvre 1 generates the offset between $\hat{x}_b$ and $\hat{v}$, i.e. an angle of attack. The rocket direction is aligned with the direction of the plane selected by the azimuth ψ_A. The angle of attack α will become zero at the time $t_v + t_p + t_{po}$, and this is the initial condition for the next phase of the launcher trajectory.

Note that the pitch manoeuvre depends on a single parameter, since the rocket's angular velocity $\dot{\theta}_0$ is a fixed parameter of the launcher and the manoeuvre depends only on the duration t_p. In alternative the angle $\theta_p = \theta_0 - \dot{\theta}_0\, t_p$ at the end of the pitch manoeuver can be taken as guidance parameter and the time duration t_p is derived by $t_p = \frac{\theta_0 - \theta_p}{\dot{\theta}_0}$.

2.4 Gravity Turn

After the pitch over, the guidance must keep the angle α at zero. This is due to the lateral load constraint on the structure created by the lift and quantified by the $q\,\alpha$ constraint (for instance $q\,\alpha_{max} = 10^6\ Pa$). Since the velocity is already high at the pitch over, the only way to reduce the $(q\,\alpha)$ term is to keep the angle of attack equal to zero thrusting along the relative velocity direction after pitch over. This is done until the rocket has reached a high enough altitude for the dynamic pressure q to decrease drastically due to the exponential decrease in air density with the altitude.

This arc of the trajectory is called gravity turn: the direction of the launcher velocity (hence the thrust direction, according to the guidance law) is changed only by the effect of gravity which turns the launcher trajectory. As there is no possibility of correcting the trajectory during this phase, it is very important to choose the right attitude angle θ_p at the end of the pitch manoeuvre, since this angle determines the entire atmospheric arc of the trajectory. Only outside the atmosphere will the launcher be able to correct its trajectory and compensate for the perturbations that occur during the flight.

The gravity turn is generally the arc with the longest duration of thrust: the rocket travels a certain path and it is important to monitor the launcher trajectory projected on the Earth surface, the ground track.

To determine the ground track it is necessary to obtain the polar coordinates of the rocket centre of mass: the distance r, the longitude λ and the latitude L. These, together with the date T_0 and time t_0 of the lift-off, define the position of the rocket centre of mass in the **Local Horizon Frame** (LHF) with axis $(\hat{r}, \hat{E}, \hat{N})$.

The absolute longitude of the rocket centre of mass is given: $\lambda_a = \lambda_{G_0} + \omega_E t_0 + \lambda$ and the LHF is related to the ECI frame by the rotations of angles λ_a and L represented by two rotation matrices:

$$R^3_{\lambda_a} = \begin{bmatrix} \cos\lambda_a & \sin\lambda_a & 0 \\ -\sin\lambda_a & \cos\lambda_a & 0 \\ 0 & 0 & 1 \end{bmatrix}$$

$$R^2_{-L} = \begin{bmatrix} \cos L & 0 & \sin L \\ 0 & 1 & 0 \\ -\sin L & 0 & \cos L \end{bmatrix}$$

so

$$\begin{bmatrix} \hat{r} \\ \hat{E} \\ \hat{N} \end{bmatrix} = R^2_{-L}\, R^3_{\lambda_a} \begin{bmatrix} \hat{c}_1 \\ \hat{c}_2 \\ \hat{c}_3 \end{bmatrix} \tag{2.23}$$

The origin of the LHF is the centre of the Earth, although the axes $(\hat{r}, \hat{E}, \hat{N})$ are sometimes shown as being attached to the rocket centre of mass to emphasize that the frame moves with the rocket centre of mass.

Since the gravity turn guidance requires the thrust to be along the relative velocity direction, the equation of motion is conveniently written in terms of the **Relative Velocity Frame** (RVF) $(\hat{\tau}, \hat{v}, \hat{h})$ having the direction $\hat{v}$ of the relative velocity as the second axis of the frame. This direction is determined by a rotation of the heading or coazimuth angle ζ_R around $\hat{r}$:

$$R^1_{\zeta_R} = \begin{bmatrix} 1 & 0 & 0 \\ 0 & \cos\zeta_R & \sin\zeta_R \\ 0 & -\sin\zeta_R & \cos\zeta_R \end{bmatrix}$$

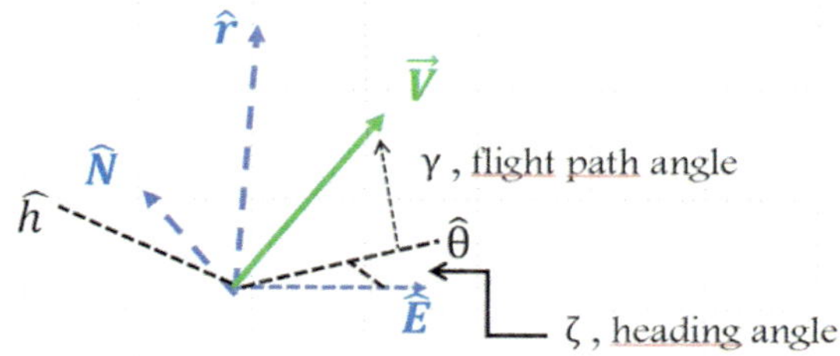

Fig. 2.5 The relative velocity frame

followed by a clockwise rotation of flight path angle γ_R around the axis $\hat{h}$:

$$R^3_{-\gamma_R} = \begin{bmatrix} \cos\gamma_R & -\sin\gamma_R & 0 \\ \sin\gamma_R & \cos\gamma_R & 0 \\ 0 & 0 & 1 \end{bmatrix}$$

so

$$\begin{bmatrix} \hat{\tau} \\ \hat{v} \\ \hat{h} \end{bmatrix} = R^3_{-\gamma_R} R^1_{\zeta_R} \begin{bmatrix} \hat{r} \\ \hat{E} \\ \hat{N} \end{bmatrix} \tag{2.24}$$

see Fig. 2.5. The following diagram (Fig. 2.6) summarises the different reference frames and their relationships.

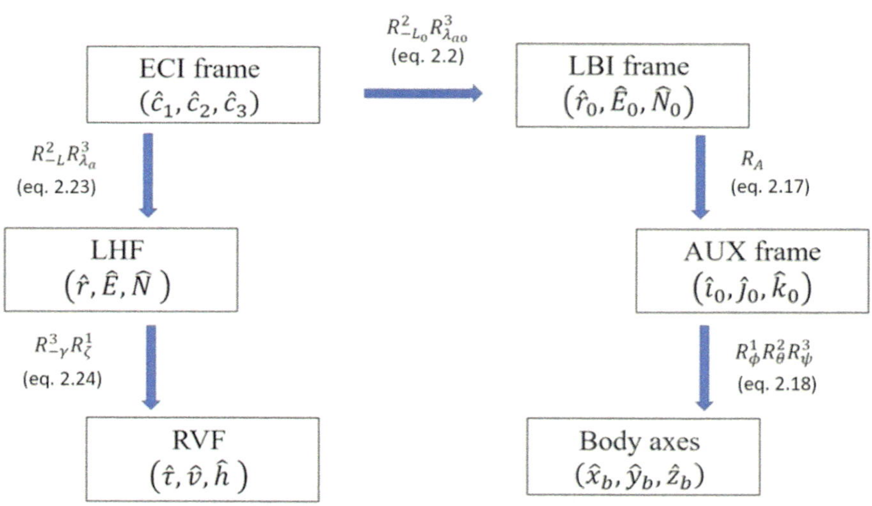

Fig. 2.6 The reference frames in the equation of flight, the rotation matrices between the frames and their related equations

The kinetic equation is written in LHF, related to ECI frame by the angular velocity[1].

$$\boldsymbol{\omega}_0 = \dot{\lambda}_a \hat{c}_3 - \dot{L}\hat{E} \quad , with \;\; \hat{c}_3 = \sin L\,\hat{r} + \cos L\,\hat{N} \tag{2.25}$$

The dynamic equation is written in the RVF, related to ECI by the angular velocity:

$$\boldsymbol{\omega} = \boldsymbol{\omega}_0 + \dot{\zeta}_R \hat{r} - \dot{\gamma}_R \hat{h} \tag{2.26}$$

The kinetic equation: In the Local Horizon Frame:

$$\frac{d\mathbf{r}}{dt} = \dot{r}\hat{r} + \boldsymbol{\omega}_0 \wedge \mathbf{r} = \mathbf{V}_R + \boldsymbol{\omega}_E \wedge \mathbf{r} \tag{2.27}$$

By (2.25) the left hand side of (2.27) corresponds to

$$\dot{r}\hat{r} + r\dot{\lambda}_a \cos L\hat{E} + r\dot{L}\hat{N} \tag{2.28}$$

Let the relative velocity V_R have $(\hat{r}, \hat{E}, \hat{N})$ components (see Fig. 2.5):

$$V_E = V_R \cos\gamma_R \cos\zeta_R \tag{2.29}$$

$$V_N = V_R \cos\gamma_R \sin\zeta_R \tag{2.30}$$

$$V_r = V_R \sin\gamma_R \tag{2.31}$$

then the right hand side of (2.27) is:

$$V_R \sin\gamma_R\,\hat{r} + V_R \cos\gamma_R \cos\zeta_R \hat{E} + V_R \cos\gamma_R \sin\zeta_R\,\hat{N} + r\omega_E \cos L\hat{E} \tag{2.32}$$

Comparing (2.28) to (2.32) one has the three kinetic scalar equations:

$$\begin{cases} \dot{r} = V_R \sin\gamma_R \\ \dot{\lambda}_a = \omega_E + \frac{V_R \cos\gamma_R \cos\zeta_R}{r\cos L} \\ \dot{L} = \frac{V_R \cos\gamma_R \sin\zeta_R}{r} \end{cases} \tag{2.33}$$

The dynamic equation is written with respect to the relative velocity frame RVF $(\hat{\tau}, \hat{\upsilon}, \hat{h})$. Then

$$\frac{d\mathbf{V}}{dt} = \frac{d}{dt}(\mathbf{V}_R + \boldsymbol{\omega}_E \wedge \mathbf{r}) = \frac{d\mathbf{V}_R}{dt} + \boldsymbol{\omega}_E \wedge \frac{d\mathbf{r}}{dt}$$

That is

$$\frac{d\mathbf{V}_R}{dt} = \dot{V}_R \hat{\upsilon} + \boldsymbol{\omega} \wedge \mathbf{V}_R + \omega_E \wedge (\dot{r}\hat{r} + \boldsymbol{\omega}_0 \wedge \mathbf{r}) \tag{2.34}$$

[1] Note that the minus sign is due to the clockwise rotation of angle L

where $\boldsymbol{\omega}$ is the angular velocity between ECI and RVF of formula (2.26) and $\boldsymbol{\omega}_0$ is the angular velocity between ECI and LH frames of formula (2.25).

The acceleration is the sum of 4 terms:

[1] $\dot{V}_R \hat{\upsilon}$
[2] $\boldsymbol{\omega} \wedge \mathbf{V}_R$
[3] $\boldsymbol{\omega}_E \wedge \underline{\dot{r}}$
[4] $\boldsymbol{\omega}_E \wedge (\boldsymbol{\omega}_0 \wedge \mathbf{r})$.

Once these terms are written in RVF the three components $(\hat{\tau}, \hat{\upsilon}, \hat{h})$ of the vector equality (2.34) provide the equations for $\dot{\gamma}_R$, $\dot{V}_R$, $\dot{\zeta}_R$ respectively, that is the equations of the dynamics.

The term [1] is already in the RVF, then the other three terms are now derived.

Derivation of term [2]

Taking into account the definitions (2.26), (2.25), this term is equal to:

$$\boldsymbol{\omega} \wedge \mathbf{V}_R = \boldsymbol{\omega}_0 \wedge \mathbf{V}_R + \dot{\zeta}_R \hat{r} \wedge \mathbf{V}_R - \dot{\gamma}_R \hat{h} \wedge \mathbf{V}_R =$$

$$= (\dot{\lambda}_a \sin L \hat{r} - \dot{L} \hat{E} + \dot{\lambda}_a \cos L \hat{N}) \wedge \mathbf{V}_R + \dot{\zeta}_R \hat{r} \wedge \mathbf{V}_R - \dot{\gamma}_R \hat{h} \wedge \mathbf{V}_R$$

Use the inverse of the relationship (2.24) to write $(\hat{r}, \hat{E}, \hat{N})$ as function of $(\hat{\tau}, \hat{\upsilon}, \hat{h})$, so that the above formula gives:

$$\begin{aligned}
&(\dot{\lambda}_a \sin L + \dot{\zeta}_R)(\cos \gamma_R \hat{\tau} + \sin \gamma_R \hat{\upsilon}) \wedge \mathbf{V}_R \\
&\quad - \dot{L}(-\sin \gamma_R \cos \zeta_R \hat{\tau} + \cos \gamma_R \cos \zeta_R \hat{\upsilon} - \sin \zeta_R \hat{h}) \wedge \mathbf{V}_R \\
&\quad + \dot{\lambda}_a \cos L(-\sin \gamma_R \sin \zeta_R \hat{\tau} + \cos \gamma_R \sin \zeta_R \hat{\upsilon} \\
&\quad + \cos \zeta_R \hat{h}) \wedge \mathbf{V}_R + \dot{\gamma} V_R \hat{\tau}
\end{aligned}$$

and performing the vector products one has:

Term [2]

$$[\dot{\gamma} V_R - V_R \dot{L} \sin \zeta_R - V_R \cos L \dot{\lambda}_a \cos \zeta_R]\hat{\tau} +$$

$$+[\dot{\zeta}_R V_R \cos \gamma_R + V_R \sin L \dot{\lambda}_a \cos \gamma_R + V_R \sin \gamma_R \cos \zeta_R \dot{L} - V_R \cos L \dot{\lambda}_a \sin \gamma_R \sin \zeta_R]\hat{h}$$

Derivation of Term [3]

Write $\hat{c}_3$ as function of $(\hat{r}, \hat{E}, \hat{N})$ (see (2.25)), then

$$\boldsymbol{\omega}_E \wedge \dot{r}\hat{r} = \dot{r}\omega_E(\sin L \hat{r} + \cos L \hat{N}) \wedge \hat{r} = \dot{r}\omega_E \cos L \, \hat{E}$$

Consider $\hat{E}$ in RVF by the inverse of (2.24). One has:

Term [3]

$$\omega_E\,\dot{r}\;cos L[-\cos\zeta_R\cos L\hat{\tau}+\cos\gamma_R\cos\zeta_R\cos L\hat{\upsilon}-\sin\zeta_R\cos L\hat{h}]$$

Derivation of term [4]

$$\boldsymbol{\omega}_E\wedge(\boldsymbol{\omega}_0\wedge\mathbf{r})$$

$$\omega_E\hat{c}_3\wedge[(\dot{\lambda}_a\sin L\hat{r}-\dot{L}\hat{E}+\dot{\lambda}_a\cos L\hat{N})\wedge\mathbf{r}]$$

$$\omega_E(\sin L\hat{r}+\cos L\hat{N})\wedge(r\dot{L}\hat{N}+r\dot{\lambda}_a\cos L\hat{E})$$

$$-r\omega_E\cos^2 L\dot{\lambda}_a\hat{r}-r\omega_E\sin L\dot{L}\hat{E}+r\omega_E\sin L\cos L\dot{\lambda}_a\hat{N}$$

Using the rotation matrix between $(\hat{r},\hat{E},\hat{N})$ and $(\hat{\tau},\hat{\upsilon},\hat{h})$ we get:

Term [4]

$$r\omega_E[-c_L^2 c_{\gamma_R}\dot{\lambda}_a+s_L s_\gamma c_{\zeta_R}\dot{L}-s_L c_L s_{\gamma_R}s_{\zeta_R}\dot{\lambda}_a]\hat{\tau}+$$

$$+r\omega_E[c_L^2 s_{\gamma_R}\dot{\lambda}_a-s_L c_{\gamma_R}c_{\gamma_R}\dot{L}+s_L c_L c_{\gamma_R}s_{\zeta_R}\dot{\lambda}_a]\hat{\upsilon}+r\omega_E[s_L c_L c_{\zeta_R}]\hat{h}$$

Now the four terms [1–4] are summed to get equations along $\hat{\tau}$, $\hat{\upsilon}$, $\hat{h}$. Namely, along τ:

$$\begin{aligned}&-r\omega_E c_L^2 r\dot{\lambda}_a c_{\gamma_R}+r\omega_E s_L\dot{L}s_{\zeta_R}c_{\gamma_R}-r\omega_E s_L c_L r\dot{\lambda}_a s_{\gamma_R}s_{\zeta_R}\\&-\omega_E V_R c_L c_{\zeta_R}+\dot{\gamma}_R V_R-V_R\dot{L}s_{\zeta_R}-V_R c_L\dot{\lambda}_a c_{\zeta_R}\end{aligned}$$

Use the kinetic equations to write the derivatives $\dot{r}$, $\dot{L}$, $\dot{\lambda}_a$ as function of V_R, γ_R and ζ_R: after some simplifications we get the

dynamical equation along $\hat{\tau}$:

$$\dot{\gamma}_R=2\omega_E\cos L\cos\zeta_R+\frac{V_R}{r}\cos\gamma_R+\frac{r\omega_E^2}{V_R}\cos L(c_L c_{\gamma_R}+s_L s_{\gamma_R}s_{\zeta_R})+\frac{F_\tau}{mV_R}\tag{2.35}$$

Considering the terms along $\hat{\upsilon}$ and using the kinetic equations one has:

dynamical equation along $\hat{\upsilon}$:

$$\dot{V}_R=r\omega_E^2\cos L(\cos L\sin\gamma_R-\sin L\cos\gamma_R\sin\zeta_R)+\frac{F_\upsilon}{m}\tag{2.36}$$

The terms along $\hat{h}$ are:

$$\begin{aligned} & r\omega_E s_L \dot{L} s_{\zeta_R} + r\omega_E s_L c_L c_{\zeta_R} \dot{\lambda}_a - \omega_E V_R s_{\gamma_R} s_{\zeta_R} c_L \\ & + V_R s_L \dot{\lambda}_a c_{\gamma_R} + \dot{\zeta}_R V_R c_{\gamma_R} + V_R s_{\gamma_R} c_{\zeta_R} \dot{L} - V_R c_L s_{\gamma_R} s_{\zeta_R} \dot{\lambda}_a \end{aligned}$$

by the kinetic equations; one has:

dynamical equation along $\hat{h}$:

$$\dot{\zeta}_R = -2\omega_E sinL - \frac{r\omega_E^2}{V_R \cos\gamma_R} c_L s_L c_{\zeta_R} + 2\omega_E c_L \tan\gamma_R s_{\zeta_R} - \frac{V_R}{r} c_{\gamma_R} \tan L c_{\zeta_R} + \frac{F_h}{m V_R \cos\gamma_R} \tag{2.37}$$

We know determine gravity, drag, lift and thrust in the RVF.

Gravity

$$\mathbf{g} = -\frac{\mu}{r^2}\hat{r} = -\frac{\mu}{r^2}(\cos\gamma_R \hat{\tau} + \sin\gamma_R \hat{\upsilon}) \tag{2.38}$$

Drag

$$\frac{\mathbf{D}}{m} = \frac{1}{2}\rho V_R^2 \frac{C_D S}{m}(-\hat{\upsilon}) \tag{2.39}$$

Lift

The lift direction is determined by a further rotation around the velocity direction $\hat{\upsilon}$ moving $\hat{\tau}$ over the lift direction $\hat{l}_a$:

$$\begin{bmatrix} \hat{l}_a \\ \hat{\upsilon} \\ \hat{h}_w \end{bmatrix} = \begin{bmatrix} \cos\sigma & 0 & -\sin\sigma \\ 0 & 1 & 0 \\ \sin\sigma & 0 & \cos\sigma \end{bmatrix} \begin{bmatrix} \hat{\tau} \\ \hat{\upsilon} \\ \hat{h} \end{bmatrix}$$

Then the lift acceleration is equal to:

$$\frac{\mathbf{L}}{m} = \frac{1}{2}\rho V_R^2 \frac{C_L S}{m} \hat{l}_a = \frac{1}{2}\rho V_R^2 \frac{C_L S}{m} (\cos\sigma\, \hat{\tau} - \sin\sigma\, \hat{h}) \tag{2.40}$$

Thrust

The thrust direction is assumed aligned with the rocket longitudinal axis and is defined by two angles α_T and β_T with respect to the RVF so that

$$\frac{\mathbf{T}}{m} = g\,n\,[\sin\alpha_T \cos\beta_T\,\hat{\tau} + \cos\alpha_T \cos\beta_T\,\hat{\upsilon} + \sin\beta_T\,\hat{h}] \tag{2.41}$$

where n is the thrust to weight ratio: $n = \frac{T}{gm}$.

In summary, the accelerations $\frac{F_\tau}{m}$, $\frac{F_\theta}{m}$, $\frac{F_h}{m}$ can be derived by adding the terms of formulas (2.38), (2.39), (2.40), (2.41):

$$\frac{F_\tau}{m} = -\frac{\mu}{r^2}\cos\gamma + \frac{1}{2}\rho V_R^2 \frac{C_L S}{m}\cos\sigma + gn\,\cos\beta_T\,\sin\alpha_T \tag{2.42}$$

$$\frac{F_\upsilon}{m} = -\frac{\mu}{r^2}\sin\gamma - \frac{1}{2}\rho V_R^2 \frac{C_D S}{m} + gn\,\cos\beta_T\,\cos\alpha_T \tag{2.43}$$

$$\frac{F_h}{m} = -\frac{1}{2}\rho V_R^2 \frac{C_L S}{m}\sin\sigma - gn\,\sin\beta_T \tag{2.44}$$

$(r, \lambda_a, L, V_R, \gamma, \xi)$ are the state variables of the kinetic and dynamical equations and the control variables are:

- The thrust intensity T.
- The angles α_T, β_T defining the direction of the thrust direction with respect to the RVF.
- The angle σ defining the out of plane component of the lift.

In gravity turn the thrust is aligned along the relative velocity direction $\hat{\upsilon}$, then the control angles are $\alpha_T = \beta_T = 0$. Note also that the zero incidence angle α implies no lift, then the dynamic equations in gravity turn are:

$$\begin{cases} \dot{V}_R = r\omega_E^2 \cos L(\cos L \sin\gamma_R - \sin L \cos\gamma_R \sin\zeta_R) - \frac{\mu}{r^2}\sin\gamma_R - \frac{1}{2}\rho V_R^2 \frac{C_D S}{m} + gn \\ \dot{\gamma}_R = 2\omega_E \cos L \cos\zeta_R + \frac{V_R}{r}\cos\gamma_R + \frac{r\omega_E^2}{V_R}\cos L(c_L c_{\gamma R} + s_L s_{\gamma R} s_{\zeta R}) - \frac{\mu}{V_R r^2}\cos\gamma_R \\ \dot{\zeta}_R = -2\omega_E sinL - \frac{r\omega_E^2}{V_R \cos\gamma_R} c_L s_L c_{\zeta R} + 2\omega_E c_L \tan\gamma_R s_{\zeta R} - \frac{V_R}{r} c_{\gamma R} \tan L c_{\zeta R} \end{cases} \tag{2.45}$$

To implement these equations after the pitch over, it is necessary to transform the final state of the pitch manoeuvre in LBI coordinates to the initial state of the gravity turn in RVF coordinates. This is done in the following steps:

- From LBI coordinates to ECI coordinates

$$\mathbf{r} = (X_1 X_2 X_3) \begin{bmatrix} \hat{r}_0 \\ \hat{E}_0 \\ \hat{N}_0 \end{bmatrix} = (X_1 X_2 X_3)\, R^3_{-\lambda_a}\, R^2_{L_0} \begin{bmatrix} \hat{c}_1 \\ \hat{c}_2 \\ \hat{c}_3 \end{bmatrix}$$

Then the ECI coordinates of **r** are:

$$(XYZ) = (X_1 X_2 X_3)\, R^3_{-\lambda_a}\, R^2_{L_0} \tag{2.46}$$

The same is for the relative velocity:

$$(V_X V_Y V_Z) = (V_1 V_2 V_3)\, R^3_{-\lambda_a}\, R^2_{L_0} \tag{2.47}$$

- From ECI to LHF coordinates: from the cartesian coordinates (XYZ) we can get the distance r, the absolute longitude λ_a and the latitude L of the rocket centre of mass (from cartesian to polar coordinates) by the relations:

$$\begin{aligned} X &= r \cos L \cos \lambda_a \\ Y &= r \cos L \sin \lambda_a \\ Z &= r \sin L \end{aligned}$$

then

$$r = \sqrt{X^2 + Y^2 + Z^2} \tag{2.48}$$

$$L = asin(\frac{Z}{r}) \tag{2.49}$$

$$\cos \lambda_a = \frac{X}{r \cos L}\, , \sin \lambda_a = \frac{Y}{r \cos L} \quad \longrightarrow \quad \lambda_a \tag{2.50}$$

Then the relative velocity V_R has LH components:

$$(V_r V_E V_N) = (V_X V_Y V_Z) R^3_{-\lambda_a}\, R^2_L$$

- From LHF to RVF coordinates: the relative velocity has LHF components that are related to the RVF components by:

$$\begin{aligned} V_r &= V_R \sin \gamma_R \\ V_E &= V_R \cos \gamma_R \cos \zeta_R \\ V_N &= V_R \cos \gamma_R \sin \zeta_R \end{aligned}$$

Then

$$V_R = \sqrt{V_r^2 + V_E^2 + V_N^2} \tag{2.51}$$

$$\gamma_R = asin\frac{V_r}{V_R} \tag{2.52}$$

$$cos\zeta_R = \frac{V_E}{V_R \cos\gamma_R}, \ \sin\zeta_R = \frac{V_N}{V_R \cos\gamma_R}, \ \longrightarrow \zeta_R \tag{2.53}$$

The transformation from LBI to RVF coordinates is completed, and the state variables corresponding to the initial values of the gravity turn equations are determined.

These equations are integrated until the conditions $(q\,\alpha < q\alpha)_{max}$ is satisfied by α small. The constraint on α disappears at high altitude (when $q \sim 0$), typically at the end of the second stage of a three stage launcher or the third stage of a four stage launcher.

2.5 Analytic Formulas for Multistage Rocket in Gravity Turn Trajectory

As mentioned above, the state of the launcher at the end of the gravity turn is very important for the success of the mission.

To determine a state at the end of the atmospheric phase suitable for reaching the target orbit, a simplified model of the gravity turn trajectory is now considered to obtain analytical formulae to be used as input for the trajectory design.

Some observations and hypotheses simplify the equation of the motion. The first point is that the duration of the atmospheric arc is limited to a few tens of seconds, so the Earth rotation can be ignored and the terms with ω_E can be set to zero in the Eq. (2.45).

The slope of the trajectory is generally high, so the range distance is limited and the flat Earth hypothesis can be introduced with a small error. Flat Earth conditions are obtained by considering the limit to infinity of the radius of curvature of the spherical Earth R_E, so that the spherical surface locally becomes the plane tangent to the surface and the gravity direction is unique and opposite to the normal to the tangent plane, see Fig. 2.7.

In formulae the flat Earth equations are obtained computing the limit $R_E \longrightarrow \infty$ in the right hand side of equations (2.33), (2.45).

Note that $r = R_E + h$ and $g = \frac{\mu}{R_E^2}$ so:

$$lim_{R_E \longrightarrow \infty}\left(\frac{V_R}{r}\cos\gamma\right) = 0$$

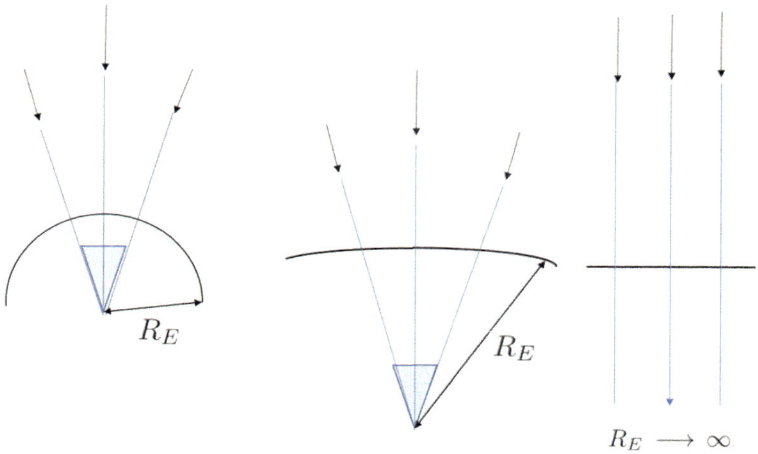

Fig. 2.7 Flat Earth as limit of $R_E \longrightarrow \infty$

$$lim_{R_E \longrightarrow \infty} \frac{\mu}{r^2} = lim_{R_E \longrightarrow \infty} \frac{g\, R_E^2}{(R_E + h)^2} = g$$

The above limits change Eqs. (2.33), (2.45) into:

$$\begin{cases} \dot{V} = -g \sin\gamma - \frac{D}{m} + g\, n \\ \dot{\gamma} = -\frac{g}{V} \cos\gamma \\ \dot{r} = V \sin\gamma \end{cases} \tag{2.54}$$

These equations have an analytic solution by two further hypothesis [12]:

- negligible drag effect
- constant thrust to mass ratio.

The second is a rather strong hypothesis which is treated by considering the average value $\bar{n}$ of the thrust to mass ratio introduced in Sect. 1.2.

The rocket has initial mass m_0 with $m_0 = m_p + m_s$ where m_p is the mass of propellant. The time dependence of the mass is $m = m_0 - m_d\, t = m_0(1 - q\, t)$ $(q = \frac{m_d}{m_0})$ and the structural mass is defined by $u = \frac{m_s}{m_0}$, so $u = 1 - q\, t_b$ where t_b is the time to burn the propellant. Then the average thrust to weight ratio is computed:

$$\bar{n} = \frac{1}{t_b} \int_0^{tb} \frac{T}{gm}\, dt = \frac{n_0}{1-u} \ln\left(\frac{1}{u}\right)$$

with n_0 thrust over initial mass ratio and $u = \frac{m_s}{m_0}$ structural mass ratio.

Now two transformations are defined:

1. From the flight path angle variable γ to the kick angle $\chi = \frac{\pi}{2} - \gamma$.
2. From the kick angle χ to the variable $z = \tan\frac{\chi}{2}$.

By the first transformation we have:

$$\begin{cases} \dot{V} = -g\,\cos\chi + g\bar{n} \\ \dot{\chi} = \frac{g}{V}\,\sin\chi \end{cases} \tag{2.55}$$

The following trigonometrical equations are recalled:

$$\begin{aligned}
\cos\chi &= \cos\left(2\,\frac{\chi}{2}\right) = \cos^2\frac{\chi}{2} - \sin^2\frac{\chi}{2} = \cos^2\frac{\chi}{2}\left(1 - \tan^2\frac{\chi}{2}\right) \\
\cos^2\chi &= \frac{1}{1+\tan^2\frac{\chi}{2}} = \frac{1}{1+z^2} \\
\cos\chi &= \frac{1-\tan^2\frac{\chi}{2}}{1+\tan^2\frac{\chi}{2}} = \frac{1-z^2}{1+z^2} \\
\sin\chi &= \sin\left(2\,\frac{\chi}{2}\right) = 2\sin\frac{\chi}{2}\cos\frac{\chi}{2} = 2\tan\frac{\chi}{2}\cos^2\frac{\chi}{2} = \frac{2z}{1+z^2}
\end{aligned}$$

showing that the z-variable allows to transform the trigonometric function into rational functions, so the Eq. (2.55) becomes:

$$\begin{cases} \dot{V} = -g\,\frac{1-z^2}{1+z^2} + g\bar{n} \\ \dot{\chi} = \frac{g}{V}\,\frac{2z}{1+z^2} \end{cases} \tag{2.56}$$

Moreover

$$z = \tan\frac{\chi}{2} \quad \Rightarrow \quad \dot{z} = \frac{1+z^2}{2}\,\dot{\chi}$$

and (2.56) is equal to:

$$\begin{cases} \dot{V} = g\,\bar{n} - g\,\dfrac{1-z^2}{1+z^2} \\ \dot{z} = \dfrac{g}{V}\,z \end{cases} \tag{2.57}$$

Dividing the first by the second equation of the system (2.57) one has:

$$\frac{dV}{dz} = \bar{n}\frac{V}{z} - \frac{V}{z}\frac{1-z^2}{1+z^2}$$

that is

$$\frac{dV}{V} = \left(\frac{\bar{n}}{z} - \frac{1}{z}\frac{1-z^2}{1+z^2}\right)dz$$

By integration:

$$\begin{aligned} log V &= \bar{n} \ln z + \ln\left(\frac{1+z^2}{z}\right) + C = \\ &= \ln z^{\bar{n}} + \ln\left(\frac{1+z^2}{z}\right) + C = \\ &= \ln\left[z^{\bar{n}-1}(1+z^2)\right] + C = \\ &= \ln\left[A\, z^{\bar{n}-1}(1+z^2)\right] \end{aligned}$$

Therefore the following analytic formula is obtained:

$$V = A\, z^{\bar{n}-1}(1+z^2) \tag{2.58}$$

The velocity in (2.58) is a function of the variable z. Its time dependence can be obtained deriving the relation between time and z. Substitution of (2.58) in the second equation of the system (2.57) gives:

$$\dot{z} = \frac{dz}{dt} = -\frac{g}{A}\frac{1}{z^{\bar{n}-2}(1+z^2)}$$

By separation of variables and integrating from $t_0 = 0$ to t one has the dependence of the time on the variable z:

$$t = \frac{A}{g}\left(\frac{z^{\bar{n}-1}}{\bar{n}-1} + \frac{z^{\bar{n}+1}}{\bar{n}+1}\right) \tag{2.59}$$

It is supposed here that the gravity turn starts from the beginning of the launch trajectory, so the initial conditions are:

$$\begin{cases} V_0 = 0 \\ \gamma_0 = \frac{\pi}{2} \\ \chi_0 = 0 \Rightarrow z_0 = 0 \end{cases}$$

Imposing these initial conditions in the solution (2.58), one has

$$V_0 = A\, z_0^{\bar{n}-1}(1+z_0^2) = 0 \quad \forall A$$

i.e. the initial conditions are verified for every value of the constant A. This means that there are infinite solutions (parameterised by A) corresponding to the same initial conditions. This violation of the uniqueness of the solution of a differential equation (Chauchy's theorem) is due to the singularity of $\dot{z}$ for $V = 0$ in the system (2.57).

One can take advantage of this multiplicity of solutions to select for example the solution that achieves a chosen flight path angle γ_f at the burn out time t_b [13]. The chosen γ_f defines a value $z_f = tan(\frac{\pi/2-\gamma_f}{2})$. By Eq. (2.59) with $t = t_b$ and $z = z_f$ one gets:

$$A = g\, t_b \left(\frac{z_f^{\bar{n}-1}}{\bar{n}-1} + \frac{z_f^{\bar{n}+1}}{\bar{n}+1} \right)^{-1} \tag{2.60}$$

and the velocity is obtained inserting this value of A in the Eq. (2.58). In particular, the velocity at burn out is:

$$V_f = A\, z_f^{\bar{n}-1}(1+z_f^2) \tag{2.61}$$

Also the altitude h and range s can be easily derived [14]. For the altitude:

$$\dot{h} = V\, \sin\gamma = V\, \cos\chi = V\, \frac{1-z^2}{1+z^2}$$

Then

$$\frac{dh}{dz} = \frac{\dot{h}}{\dot{z}} = \frac{V^2}{g\, z}\, \frac{1-z^2}{1+z^2}$$

By formula (2.58) one has:

$$\frac{dh}{dz} = \frac{A^2}{g}\, \left(z^{2\bar{n}-3} - z^{2\bar{n}+1}\right)$$

The elementary integration of the above equation between current z and initial z_0 gives the formula for the altitude:

$$h(z) = h(z_0) + \frac{A^2}{g} \left[\frac{z^{2\bar{n}-2}}{2\bar{n}-2} - \frac{z^{2\bar{n}+2}}{2\bar{n}+2} \right]_{z_0}^{z_f} \tag{2.62}$$

For the range s:

$$\dot{s} = V\, \cos\gamma = V\, sin\chi = 2\, V \frac{z}{1+z^2}$$

then

$$\frac{ds}{dz} = \frac{\dot{s}}{\dot{z}} = 2\frac{A^2}{g}\, \left(z^{2\bar{n}-1} + z^{2\bar{n}+1}\right)$$

The integration gives the formula for the range:

$$s(z) = s(z_0) + 2\frac{A^2}{g} \left[\frac{z^{2\bar{n}}}{2\bar{n}} + \frac{z^{2\bar{n}+2}}{2\bar{n}+2} \right]_{z_0}^{z_f} \tag{2.63}$$

Equations (2.58), (2.62), (2.63), together with (2.60), (2.59) give the state variables of the gravity turn trajectory of a single stage rocket with flight path angle γ_f at burn out time t_b [15].

Note that space missions require multi-stage rockets then it is necessary to generalize the above formulae to any number of stages [13].

Let us start with two stages having the average parameters n_1, n_2 and burn out times t_{b_1}, t_{b_2} (Fig. 2.8).

The first stage has variables z_1, $V_1 = A_1\, z_1^{\bar{n}_1-1}(1+z_1^2)$ in the time interval $[t_{10} = 0, t_{1f} = t_{b_1}]$. The second stage has variables z_2, $V_2 = A_2\, z_2^{\bar{n}_2-1}(1+z_2^2)$ in the time interval $[t_{20} = t_{b_1}, t_{2f} = t_{b_2} + t_{b_1}]$ and the final value z_{2_f} is fixed.

Of course the two solutions must match at the boundary point:

$$z_{1f} = z_{20}$$

$$V_{1f} = V_1(z_{1f}) = V_2(z_{20}) = V_{20}$$

The two matching conditions imply

$$\frac{A_2}{A_1} = z_{20}^{\bar{n}_2-\bar{n}_1} \tag{2.64}$$

Equation 2.59 gives at first stage burn out

$$t_{b1} = \frac{A_1}{g}\left(\frac{z_{20}^{\bar{n}_1-1}}{\bar{n}_1-1} + \frac{z_{20}^{\bar{n}_1+1}}{\bar{n}_1+1}\right) =^{def} \frac{A_1}{g} f_1(z_{20}) \tag{2.65}$$

At the burn out of the second stage one has z_{2f} and

$$t_{b2} = \frac{A_2}{g}\left(\frac{z_{2f}^{\bar{n}_2-1}}{\bar{n}_2-1} + \frac{z_{2f}^{\bar{n}_2+1}}{\bar{n}_1+1} - \frac{z_{20}^{\bar{n}_2-1}}{\bar{n}_2-1} + \frac{z_{20}^{\bar{n}_2+1}}{\bar{n}_1+1}\right) =^{def} \frac{A_2}{g} f_2(z_{20}) \tag{2.66}$$

We have a system of three Eqs. (2.64), (2.65), (2.66) in the three unknowns A_1, A_2 and z_{20}.

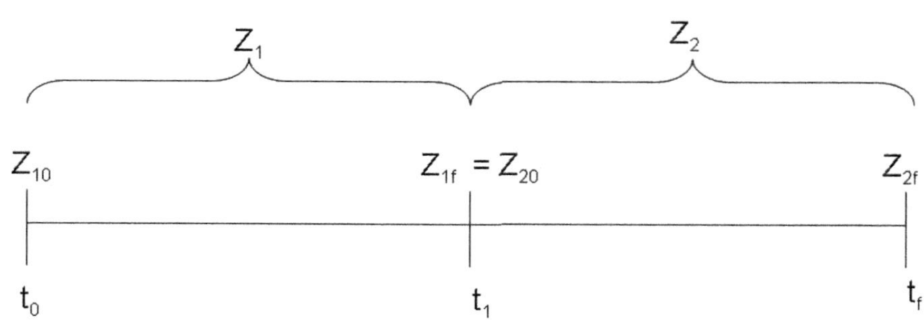

Fig. 2.8 Two stages: the z variables and the matching condition

To solve it, let us consider the ratio (use also (2.64)):

$$\frac{t_{b2}}{t_{b1}} = z_{20}^{\bar{n}_2-\bar{n}_1}\,\frac{f_2}{f_1}$$

and introduce the function of z_{20}:

$$f(z_{20}) = \frac{t_{b2}}{t_{b1}} - z_{20}^{\bar{n}_2-\bar{n}_1}\,\frac{f_2}{f_1}$$

The solution z_{20} of $f(z_{20}) = 0$ (in the range [0,1]) provides the value of the flight path angle at the end of the first stage. Moreover the Eq. (2.66) gives A_2 and Eq. (2.65) gives A_1. With these solutions the velocity profile of the two stage launcher having prescribed flight path angle z_{2f} at second stage burn out is equal to:

$$\begin{cases} V(z) = A_1\, z_1^{\bar{n}_1-1}(1+z_1^2)\ for\ z \in [0, z_{20}] \\ V(z) = A_2\, z_2^{\bar{n}_2-1}(1+z_2^2)\ for\ z \in [z_{20}, z_{2f}] \end{cases} \tag{2.67}$$

In the particular case $z_{2f} = 1$, that is zero flight path angle at the second stage burn out, we have

$$V_{2f} = 2\,A_2 \tag{2.68}$$

With same arguments one gets the altitude and range for a two stage rocket:

$$\begin{cases} h(z) = \dfrac{A_1^2}{g}\left[\dfrac{z_1^{2\bar{n}_1-2}}{2\bar{n}_1-2} - \dfrac{z_1^{2\bar{n}_1+2}}{\bar{n}_1+2}\right],\ for\ z \in [0, z_{20}] \\ h(z) = h(z_{20}) + \dfrac{A_2^2}{g}\left[\dfrac{z_2^{2\bar{n}_2-2}}{2\bar{n}_2-2} - \dfrac{z_2^{2\bar{n}_2+2}}{\bar{n}_2+2}\right]_{z_{20}}^{z},\ for\ z \in [z_{20}, z_{2f}] \end{cases} \tag{2.69}$$

$$\begin{cases} s(z) = 2\dfrac{A_1^2}{g}\left[\dfrac{z^{2\bar{n}_1}}{2\bar{n}_1} + \dfrac{z^{2\bar{n}_1+2}}{2\bar{n}_1+2}\right],\ for\ z \in [0, z_{20}] \\ s(z) = s(z_{20}) + 2\dfrac{A_2^2}{g}\left[\dfrac{z^{2\bar{n}_2}}{2\bar{n}_2} + \dfrac{z^{2\bar{n}_2+2}}{2\bar{n}_2+2}\right]_{z_{20}}^{z},\ for\ z \in [z_{20}, z_{2f}] \end{cases} \tag{2.70}$$

Any number of stages N can be handled: of course more work is needed to find out the unknown variables. Namely the N–1 matching conditions on the flight path angle generate the following N–1 matching conditions on the velocity:

$$
\begin{aligned}
V_1(z_{20}, A_1) &= V_2(z_{20}, A_2)\\
V_2(z_{30}, A_2) &= V_3(z_{30}, A_3)\\
&\vdots\\
V_{N-2}(z_{N-2\,0}, A_{N-2}) &= V_{N-1}(z_{N-1\,0}, A_{N-1})
\end{aligned}
$$

These, together with the N relationships related to each burn time

$$
\begin{aligned}
t_{b_1} &= t_{b_1}(A_1, z_{20})\\
t_{b_2} &= t_{b_2}(A_2, z_{20}, z_{30})\\
&\vdots\\
t_{b_N} &= t_{b_N}(A_N, z_{20}, \ldots, z_{N\,0})
\end{aligned}
$$

generates 2N–1 equations to be solved with respect to the 2N–1 unknowns $z_{20}, z_{30}, \ldots z_{N0}, A_1, \ldots A_N$.

By these solutions, the velocity, altitude and range profiles can be computed in the ranges $[0, z_{20}], \ldots [z_{N0}, z_f]$.

For instance for a **three stage launcher** one has:

$$t_{b_1} = \frac{A_1}{g}\left(\frac{z_{20}^{n_1-1}}{n_1-1} + \frac{z_{20}^{n_1+1}}{n_1+1}\right) \tag{2.71}$$

$$t_{b_2} = \frac{A_2}{g}\left[\left(\frac{z_{30}^{n_2-1}}{n_2-1} + \frac{z_{30}^{n_2+1}}{n_2+1}\right) - \left(\frac{z_{20}^{n_2-1}}{n_2-1} + \frac{z_{20}^{n_2+1}}{n_2+1}\right)\right] \tag{2.72}$$

$$t_{b_3} = \frac{A_3}{g}\left[\left(\frac{z_3^{n_3-1}}{n_3-1} + \frac{z_3^{n_3+1}}{n_3+1}\right)\right]_{z_{3_0}}^{z_{3_f}} \tag{2.73}$$

$$A_1 = A_2 z_{20}^{\overline{n}_2-\overline{n}_1} \tag{2.74}$$

$$A_2 = A_3\, z_{30}^{\overline{n}_3-\overline{n}_2} \tag{2.75}$$

five Eqs. (2.71)–(2.75) in the five unknowns: $A1, A_2, A_3, z_{20}, z_{30}$).

One solution is to write the variables A_1 and A_2 as a function of A_3 using the Eqs. (2.74), (2.75). Then the ratios $\frac{t_{b_2}}{t_{b_1}}, \frac{t_{b_3}}{t_{b_1}}$ are obtained by dividing (2.72), (2.73) by the Eq. 2.71. The two ratios give two equations for the two unknowns z_{20}, z_{30}. The solutions can be found within the interval [0,1], and they determine the values of the flight path angles at the end of the first and second stages. The values A_1, A_2, A_3 are derived by the Eqs. (2.71), (2.73) and the velocities at the end of each stage are obtained.

For a **four stage launcher** there are 7 equations in the 7 unknowns: $A1$, A_2, A_3, A_4, z_{20}, z_{30}, z_{40}:

$$A_1 = A_2 z_{20}^{\overline{n}_2 - \overline{n}_1} \tag{2.76}$$

$$A_2 = A_3\, z_{30}^{\overline{n}_3 - \overline{n}_2} \tag{2.77}$$

$$A_3 = A_4\, z_{40}^{\overline{n}_4 - \overline{n}_3} \tag{2.78}$$

$$t_{b_1} = \frac{A_1}{g}\left(\frac{z_{20}^{n_1-1}}{n_1-1} + \frac{z_{20}^{n_1+1}}{n_1+1}\right) \tag{2.79}$$

$$t_{b_2} = \frac{A_2}{g}\left[\left(\frac{z_{30}^{n_2-1}}{n_2-1} + \frac{z_{30}^{n_2+1}}{n_2+1}\right) - \left(\frac{z_{20}^{n_2-1}}{n_2-1} + \frac{z_{20}^{n_2+1}}{n_2+1}\right)\right] \tag{2.80}$$

$$t_{b_3} = \frac{A_3}{g}\left[\left(\frac{z_{40}^{n_3-1}}{n_3-1} + \frac{z_{40}^{n_3+1}}{n_3+1}\right) - \left(\frac{z_{30}^{n_3-1}}{n_3-1} + \frac{z_{30}^{n_3+1}}{n_3+1}\right)\right] \tag{2.81}$$

$$t_{b_4} = \frac{A_4}{g}\left[\frac{2\,n_4}{n_4^2-1} - \left(\frac{z_{40}^{n_4-1}}{n_4-1} + \frac{z_{40}^{n_4+1}}{n_4+1}\right)\right] \tag{2.82}$$

The first three equations allow to write A_1, A_2, A_3 as function of A_4 only. Then the three ratios $\frac{t_{b_2}}{t_{b_1}}$, $\frac{t_{b_3}}{t_{b_1}}$, $\frac{t_{b_4}}{t_{b_1}}$ are functions of the three unknowns z_{20}, z_{30}, z_{40}.

2.6 Equation of Losses and Generalization of the Tziolkowski Formula

A first estimate of the capability of a launcher to reach a given orbit is obtained by comparing the variation of velocity that the launcher engines can provide to the payload with the variation of velocity needed to reach the target orbital velocity. The velocity provided by the launcher depends on the characteristics of the launcher's engines and on the trajectory being followed. The popularity of the Tziolkowski formula recalled in Sect. 1.2 is due to its simplicity: only the thrust acceleration is considered and no knowledge of the launcher trajectory is required. In fact the Tziolkowski formula is obtained integrating the dynamics involving only the thrust:

$$\dot{V} = \frac{T}{m} = gn$$

Table 2.1 Tziolkowski velocity and losses $[\frac{m}{s}]$ al first stage burn out of the Vega launcher in nominal trajectory

ΔV_{Tz}	ΔV_g	ΔV_d	ΔV_α	ΔV_{cp}
3054	683	107	19	105

However the above is not the acceleration experienced by the launcher in reality: this is

$$\dot{V} = \frac{T}{m}\cos\alpha_T - \frac{\mu}{r^2}\sin\gamma_R - \frac{D}{m}$$

(see Eq. (2.36) with $\omega_E = 0$ and planar case $\zeta_R = \dot{\zeta}_R = 0$).

The variation of velocity after a boost of duration t_b is obtained by integrating the above formula, that is:

$$\Delta V = -\int_{t_0}^{t_b} \frac{\mu}{r^2}\sin\gamma\, dt - \int_{t_0}^{t_b} \frac{D}{m}\, dt + \int_{t_0}^{t_b} \frac{T}{m}\cos\alpha\, dt$$

corresponding to the **the equation of the losses**:

$$\Delta V = \int_{t_0}^{t_f} \frac{T}{m}\, dt - \int_{t_0}^{t_f} \frac{\mu}{r^2}\sin\gamma\, dt - \int_{t_0}^{t_f} \frac{D}{m}\, dt - \int_{t_0}^{t_f} \frac{T}{m}(1-\cos\alpha)dt \tag{2.83}$$

In Sect. 1.2 it is proved that the first integral is equal to the Tziolkowski variation of velocity, the other integrals are defined as:

Gravity loss: $\Delta V_g = \int_{t_0}^{t_f} \frac{\mu}{r^2}\sin\gamma\, dt$

Drag loss: $\Delta V_d = \int_{t_0}^{t_f} \frac{D}{m}\, dt$

Misalignment loss: $\Delta V_\alpha = \int_{t_0}^{t_f} \frac{T}{m}(1-\cos\alpha)\, dt$

The latter occurs when the thrust is not directed as the velocity. The Eq. (2.83) is equal to:

$$\Delta V = \Delta V_{Tz} - \Delta V_g - \Delta V_d - \Delta V_\alpha$$

i.e. the real variation of velocity is equal to the Tziolkoski term minus the "losses".

Table 2.1 gives the Tziolkowski velocity and losses for the first stage of the Vega launcher on a typical mission.

The data in the table are obtained by inserting in (2.83) the actual trajectory data of the first stage of the Vega launcher.

In the table ΔV_{cp} represents the over-pressure loss due to the fact that the nozzle is not adapted to the external pressure. Note that the Tziolkovski variation of velocity is reduced by almost 30%. This is a further evidence that the Tziolkowki formula gives an overly optimistic estimate of the velocity that can be achieved by a launcher and that the sum of the Tziolkowki variations of velocity produced by the engines of a launcher cannot give a reliable estimate of the ability to achieve the required orbital velocity.

The data in the Table also show that during the first stage the over-pressure loss is of the same order of magnitude as the drag loss and the misalignment loss is almost zero due to the gravity turn guidance; the gravitational loss is by far the largest. It is therefore important to have an estimate of the gravitational loss to get a more reliable estimate of the variation of velocity that a launcher can provide.

It should be noted that the losses depend on the trajectory data, so it is very difficult to give an a priori estimate of them. A rough estimate of the gravitational loss is given by

$$\Delta V_g = -g \int_{t_0}^{t_f} \sin\gamma \, dt \leq -g \int_{t_0}^{t_f} dt = g \left(t_f - t_0\right)$$

However this formula over-estimates the gravitational loss: for instance its value for the Vega launcher is $\Delta V_g = 1050\,\text{m/s}$ instead of 683 m/s.

For a more accurate evaluation of the gravity loss, recall that the analytical formulae obtained in the previous section provide with the velocity of the launcher under the effect of both thrust and gravity accelerations. This allows the gravity loss to be calculated using a simple analytical formula.

For instance, let us consider initial lift off condition ($V_0 = 0,\ \chi_0 = 0,\ h_0 = 0,\ R_0 = 0$) and choose the final condition $\gamma_f = 0$: this condition corresponds to $z_f = 1$. Then the formula (2.60) gives

$$A = g\, t_b \, \frac{\bar{n}^2 - 1}{2\bar{n}}$$

By formulas (2.58), (2.62), (2.63) the velocity, altitude and range at burn out are equal to:

$$V_f = 2A = g\, t_b \, \frac{\bar{n}^2 - 1}{\bar{n}} \tag{2.84}$$

$$h_f = \frac{A^2}{g} \frac{1}{\bar{n} - 1} = \frac{g t_b^2}{2} \frac{\bar{n}^2 - 1}{\bar{n}^2}$$

$$s_f = \frac{A^2}{g} \left(\frac{1}{\bar{n}} + \frac{1}{\bar{n} + 1}\right) = \frac{g t_b^2}{2} \left(\frac{\bar{n}^2 - 1}{\bar{n}}\right)^2 \left[\frac{1}{2\bar{n}} + \frac{1}{2\bar{n} + 2}\right]$$

In the present setting, the Tziolkowski formula corresponds to:

$$\Delta V_{Tz} = \int_{t_0}^{t_f} \frac{T}{m} \, dt = g \int_{t_0}^{t_f} \bar{n} \, dt = g\, \bar{n} t_b \tag{2.85}$$

The difference between (2.85) and (2.84) gives an estimate of the gravitational loss:

$$\Delta V_g = V_{Tz} - V_f = \frac{g\, t_b}{\bar{n}} \tag{2.86}$$

The gravitational losses for multistage launchers are obtained in the same way. For example consider a two stage launcher with parameters $\bar{n}_1$, $\bar{n}_2$ and t_{b_1}, t_{b_2}. To evaluate the gravitational loss we shall assume that the second stage has a zero flight path angle at the end of the burn out time. Then its final velocity V_{2f} is equal to $2A_2$.

The gravitational loss is equal to the difference between the sum of the Tziolkoski variation of velocity provided by the engines of the two stages and the final velocity $V_{2f} = 2A_2$:

$$\Delta V_g = g\,\bar{n}_1\, t_{b_1} + g\,\bar{n}_2\, t_{b_2} - 2\, A_2$$

For launchers with k-stages the gravity loss is equal to:

$$\Delta V_g = g\,\Sigma_k \bar{n}_k\, t_{b_k} - 2A_k \tag{2.87}$$

It should be noted that the formula for the final velocity of a k-stage launcher

$$V_{k_f} = 2A_k \tag{2.88}$$

includes the thrust and the gravity effect, then it improves the Tziolkowski formula in approximating the velocity variation that a launcher can provide for a given payload mass.

Of course if the velocity (2.88) is less than the required orbital velocity, the mission cannot be accomplished.

The Figs. 2.9, 2.10, 2.11 compare the analytic formulae with the numerical integration of the trajectory performed by the three stages of the Scout launcher, according to Problems 2.1, 2.2, 2.3.

The comparison between the analytical and the numerical results is fairly satisfactory, but there are differences which can be explained as follows. Figure 2.9 shows that the analytical formula overestimates the velocity because it ignores the drag effect. Moreover there is a difference between the two profiles also due to model of the thrust, where average values are assumed in the analytical formulae while accurate representation is taken in the numerical integration by interpolation in time of the real thrust data.

The difference in altitude of Fig. 2.10 is due to the difference in the velocities. The numerical derivation of the flight path angle in Fig. 2.11 shows angle at constant value close to 90 deg during the vertical arc and its variation during the pitch manoeuvre, whereas the analytic flight path angle decreases monotonically.

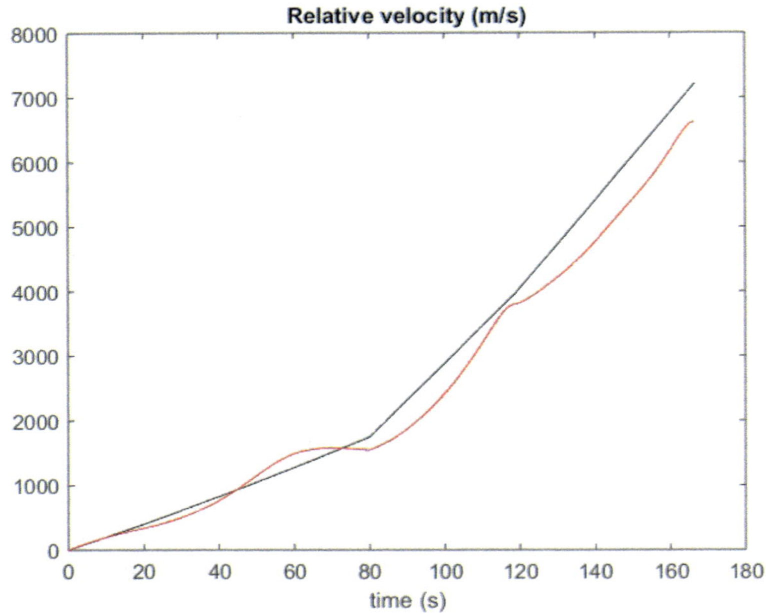

Fig. 2.9 Velocity profiles obtained numerically and analytically of the first three stages of the Scout along the trajectory defined in the Problems 2.1–2.3

2.7 Fast Evaluation of Launcher Trajectories in Atmospheric Arc

The above analytical formulae can be used to determine the state variables of a k-stage launcher at the exit of the atmospheric arc. The k-stage launcher has the average parameters $\bar{n}_1, \ldots, \bar{n}_k$ relative to a given payload mass m_u and burn out times $t_{b_1}, \ldots, t_{b_k}$. The exit of the atmospheric arc corresponds to the end of the gravity turn arc, generally occurring at the burn out of the k–1-stage of the rocket, and depends on a single parameter, the final flight path angle γ_{ex}.

In fact, as the flight path angles γ_{ex} at the k–1 stage burn out are considered in a range of values, the corresponding values $z_f = tan(\frac{\frac{\pi}{2} - \gamma_{ex}}{2})$ are computed. These values are inserted in the Eqs. (2.61)–(2.82) depending on the number of stages. For the Scout launcher, velocity and altitude at third stage burn out are plotted in Fig. 2.12 for γ_{ex} in the range $[0°, 40°]$, see Problem 2.5.

Note that for higher values of γ_{ex} we have steepest trajectories with greater gravity loss, hence reduced velocity but higher altitudes.

In the numerical and accurate simulation of the launch trajectory, the crucial parameter γ_{ex} can be related to the duration t_p of the pitch manoeuvre. Since the duration of the vertical arc t_v is fixed for safety reasons, the different values of t_p

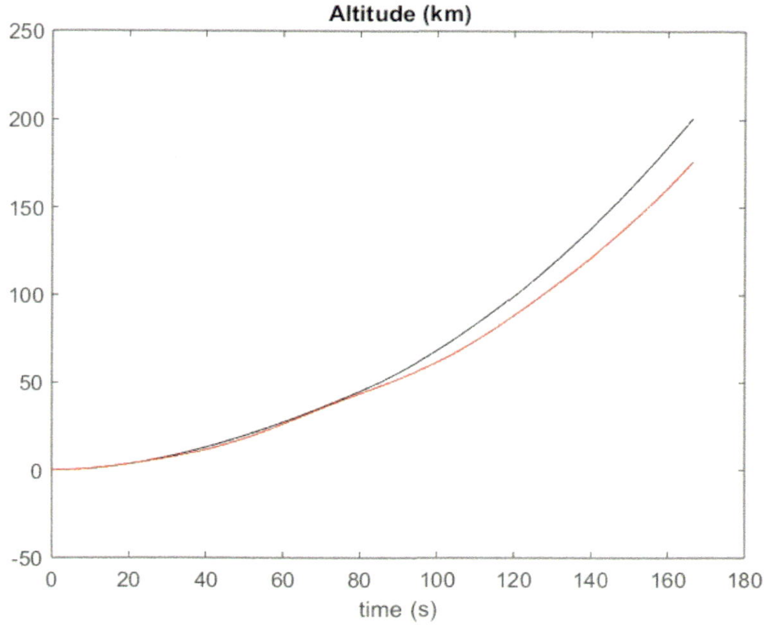

Fig. 2.10 Altitude profiles obtained numerically and analytically of the first three stages of the Scout along the trajectory defined in the Problems 2.1–2.3

produce different pitch over times, and hence different initial (and final) conditions of the gravity turn arc.

Figure 2.13 shows the relation between t_p and the flight path angle γ_{ex} for the Scout launcher, see Problem 2.6.

The numerical values of the altitudes and (relative) velocities corresponding to γ_{ex} are given in Fig. 2.14. This output can be compared with the analytical results of Fig. 2.12. The error in altitude of the analytical output is at most of the 5%, the error in velocity is of 13%.

Problems

2.1 Given models of mass with payload and adapter equal to 150 kg, thrust, drag and lift coefficients of the Scout launch vehicle (see Sects. 1.3, 1.5, 1.8) and the atmospheric density model of Sect. 1.8, determine the state variables of the rocket at the end of a vertical arc of duration $t_v = 6s$, kick angle 0.5 deg, lift-off from the San Marco platform of longitude/latitude $\lambda_0 = 40.20$ deg East, $L_0 = 2.93$ deg South and launch date T_0: 22nd October 2026 at time $t_0 = 0 : 00$.

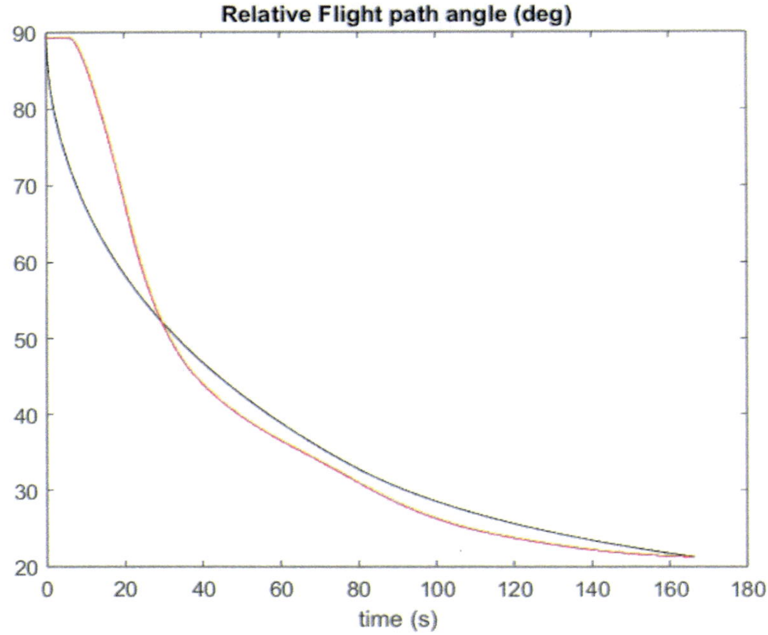

Fig. 2.11 Flight path angle profiles obtained numerically and analytically of the first three stages of the Scout along the trajectory defined in the Problems 2.1–2.3

2.2 Starting from the final conditions of the vertical launch determined in Problem 2.1, compute the pitch over time t_{po} and state variables at t_{po}, assuming that the rocket angular rate is $\dot{\theta} = 3$ deg/s, the duration of the pitch rotation is $t_p = 15$ s and the azimuth angle is $\psi_A = 132.7$ deg.

2.3 Using the state variables from Problem 2.2 compute the state variables at the end of the third stage boost with gravity turn guidance and no coasting between the stages.

2.4 Using the analytical formula (2.88), calculate the variation of velocity provided by the first three stages of the Scout and Vega launchers in gravity turn trajectory having zero flight path angle at the end of the third stage. Compare the results with the variation of velocity obtained using the Tziolkowski formula and evaluate the gravity losses for each stage.

2.5 According to the analytical formulae, estimate the velocity and altitude reached by the Scout G launcher at the burn out of the third stage with payload mass 150 kg for the range $[0°, 40°]$ of final flight path angles γ_{ex}.

2.6 Using the numerical model of Problem 2.1, determine the pitch over times t_{po} producing the flight path angles $\gamma_{ex} \in [0°, 40°]$. Determine the altitude and velocity

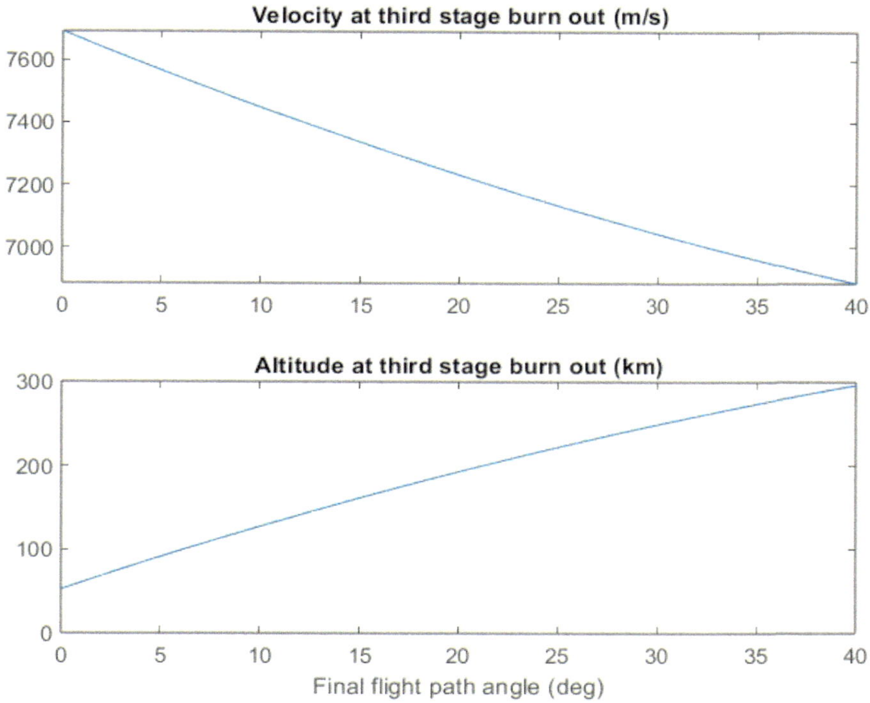

Fig. 2.12 Velocity and altitude (analytic values) achieved by the third stage of Scout G in gravity turn at third stage burn out for different final flight path angles γ_{ex}

achieved at the third stage burn out and compare these numerical results with the analytical results of Problem 2.5.

Solutions

Problems of This Chapter

2.1 The equations of motion to be solved numerically are (2.7), (2.8) with the initial conditions given in the problem. The accelerations are in (2.10), (2.11), (2.13) with $\mu = 398601\,\text{Km}^3/\text{s}^2$ and the mass m and aerodynamic coefficients modelled as it follows.

The structural mass of each stage of Scout G is equal to

$$m_{s_1} = 2030\,\text{kg},\ m_{s_2} = 695\,\text{kg},\ m_{s_3} = 98\,\text{kg},\ m_{s_4} = 28\,\text{kg}$$

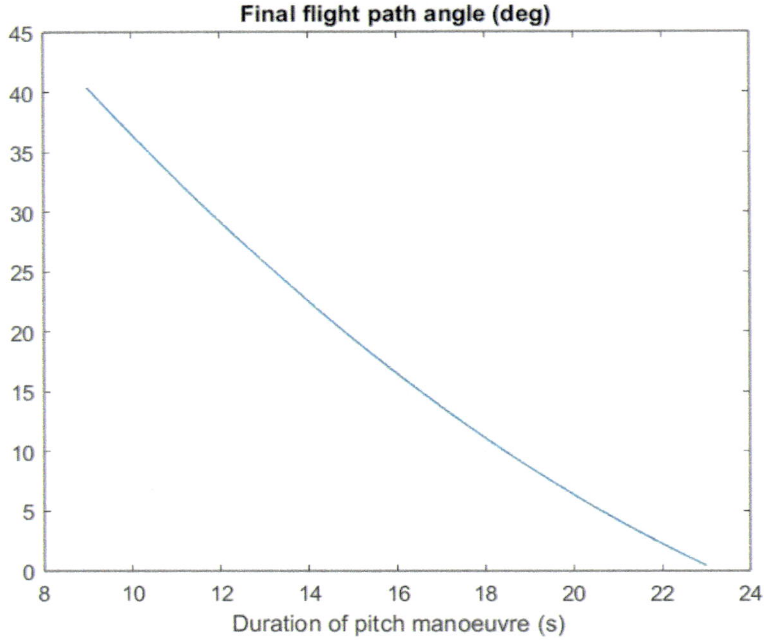

Fig. 2.13 The flight path angles γ_{ex} at the end of the third stage in gravity turn corresponding to different duration of the pitch manoeuvre t_p

The initial propellant mass of each stage is

$$m_{p_1} = 12817\,\text{kg}, m_{p_2} = 3729\,\text{kg}, m_{p_3} = 1286\,\text{kg}, m_{p_4} = 286\,\text{kg}$$

Heat shield mass $mhs = 120\,\text{kg}$ and payload mass, including adapter $mpay = 150\,\text{kg}$; then the launcher initial mass is

$$M_1 = ms1 + mp1 + ms2 + mp2 + ms3 + mp3 + ms4 + mp4 + mhs + mpay = 21239\,\text{kg}$$

This mass decreases with the propellant consumption of the first stage. The actual time variation of the first stage propellant mass is interpolated by a 6th degree polynomial:

$$m_p = p_1 t^6 + p_2 t^5 + p_3 t^4 + p_4 t^3 + p_5 t^2 + p_6 t + p_7 \,, \quad t \in [0, t_{b1}]$$

with coefficients:

$$p_1 = -4.6698e - 07,\ p_2 = 0.0001055,\ p_3 = -0.0079754,$$
$$p_4 = 0.225,\ p_5 = -1.2497,\ p_6 = -208.0075,\ p_7 = 12817.0264$$

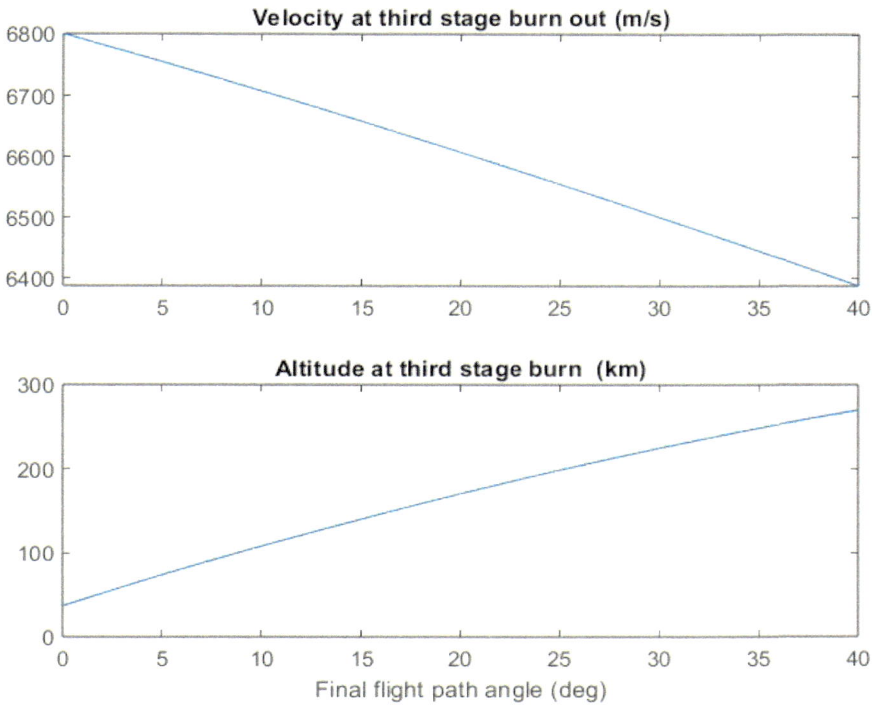

Fig. 2.14 Velocity and altitude (numerical values) achieved by the third stage of SCOUT G in gravity turn at third stage burn out for different final flight path angles γ_{ex}

The time variation of the first stage thrust T in KN, see Fig. 1.16, is interpolated by two polynomials, one in the interval $t \in [0, t_1]$, $t_1 = 57.5$ s

$$T = q_{14}\, t^4 + q_{13}\, t^3 + q_{12}^2\, t^2 + q_{11}\, t + q_{10}$$
$$q_{14} = -0.00036044, q_{13} = 0.024861, q_{12} = 0.0052959, q_{11} = -19.3722, q_{10} = 684.4673$$

The second is a polynomial of degree 4, defined in $t \in [t_1, t_{b_1}]$, $t_{b_1} = 80$ s

$$T = q_{113}\, t^3 + q_{112}\, t^2 + q_{111}^2\, t + q_{110}$$
$$q_{113} = -0.040287, q_{112} = 9.2813, q_{111} = -716.4398, q_{110} = 18541.2815$$

The aerodynamic coefficient CD_0 as function of Mach (Fig. 1.26) is interpolated in two intervals:

For $Mach \leq 1$

$$CD_0 = ma_{p1}\, Mach^4 + ma_{p2}\, Mach^3 + ma_{p3}\, Mach^2 + ma_{p4}\, Mach + ma_{p5}$$
$$ma_{p1} = 2.1333, ma_{p2} = -2.88, ma_{p3} = 1.5467, ma_{p4} = -0.2, ma_{p5} = 0.4$$

Table 2.2 Data at end of vertical arc

time (s)	Event	Altitude (m)	Rel. Vel (m/s)	Rel. fpa (deg)
0	Lift off	0	0	90
6	End vertical	217	124	89.3

For $Mach > 1$

$$CD_0 = ma_{p11}\,Mach^5 + ma_{p22}\,Mach^4 + ma_{p33}\,Mach^3 + ma_{p44}\,Mach^2 + ma_{p55}\,Mach + ma_{p66}$$
$$ma_{p11} = 0.0043776, ma_{p22} = -0.076011, ma_{p33} = 0.49389,$$
$$ma_{p44} = -1.4382, ma_{p55} = 1.6358, ma_{p66} = 0.38341$$

The aerodynamic coefficients CL_α, k_D as function of Mach (see Figs. 1.25, 1.27) are interpolated as follows:

For $Mach \le 1$

$$C_{L_\alpha} = p_1\,Mach^4 + p_2\,Mach^3 + p_3\,Mach^2 + p_4\,Mach + p_5$$
$$p_1 = 0.10074,\; p_2 = 0,\; p_3 = -0.066296,\; p_4 = 0.060556,\; p_5 = 0.25$$
$$k_D = k_1\,Mach^3 + k_2\,Mach^2 + k_3\,Mach + p_4$$
$$k_1 = -1.3333,\; k_2 = 0.85714,\; k_3 = -0.57381,\; k_4 = 4.0021$$

For $Mach \in (1, 5]$

$$C_{L_\alpha} = p_{11}\,Mach^4 + p_{22}\,Mach^3 + p_{33}\,Mach^2 + p_{44}\,Mach + p_{55}$$
$$p_{11} = 0.0037051,\; p_{22} = -0.053012,\; p_{33} = 0.28196,\; p_{44} = -0.67971,\; p_{55} = 0.79061$$
$$k_D = k_{11}\,Mach^3 + k_{22}\,Mach^2 + k_{33}\,Mach + k_{44}$$
$$k_{11} = 0.038269,\; k_{22} = -0.6749,\; k_{33} = 4.0097,\; k_{44} = -0.45598$$

For $Mach > 5$ the value at Mach = 5 is assumed.

The surface area related to the aerodynamic coefficients is $S = 0.47\,\text{m}^2$, and CL_α must be multiplied by the angle of attack α in degrees (Table 2.2).

The launch initial conditions in LBI frame (relative velocity) are

$$(X1_0, X2_0, X3_0, VX1_0, VX2_0, VX3_0) = (R_E, 0, 0, 0, 0, 0)$$

The guidance law is (2.14). At final time $t_v = 6$ s, that is the end of the vertical arc, one has:

$$\begin{aligned}(X_1, X_2, X_3) &= (6378.40\,10^3\,m, 2.78\,10^3\,m, -3\,m)\\ (V_{X1}, V_{X2}, V_{X3}) &= (124\,\text{m/s}, 1.8\,\text{m/s}, -1.0\,\text{m/s})\end{aligned} \tag{2.89}$$

corresponding to the data in Table 2.2, with relative velocity and flight path angle obtained by formulae (2.46)–(2.52)

2.2 Start with the conditions of Problem 2.1: state conditions (2.89) at time $t = t_v$, remaining propellant $m_p = 11563$ kg, then total mass at beginning of the pitch

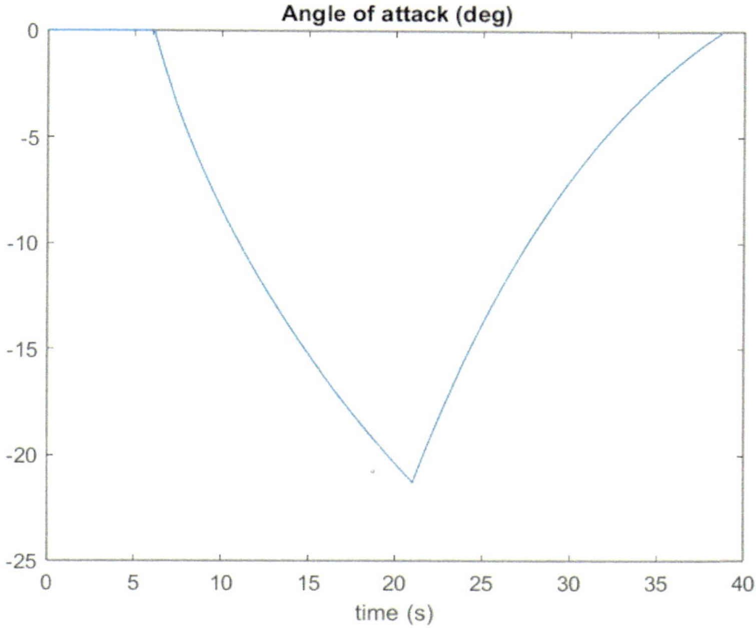

Fig. 2.15 The angle of attack α (deg) from lift off to pitch over

manoeuvre $M_1 = 19985\,\text{kg}$. Integrate the same equations from Problem 2.1 with the thrust direction from formula (2.22) with $\psi = \psi_A - \frac{\pi}{2}$ and θ is chosen according to (2.20), (2.21), pitch duration $t_p = 15$ s and pitch rate $\dot{\theta}_p = 3\,\text{deg/s}$.

The pitch manoeuvre produces a variation in the angle of attack $\alpha = \theta - \gamma_R$ as shown in the Fig. 2.15. Note that α is zero during the vertical arc and decreases during the manoeuvre up to the value $-21\,\text{deg}$. The angle then returns to zero and the pitch over time is equal to $t_{po} = 39\,\text{s}$ (counted from lift-off time $t_0 = 0$) (Table 2.3).

At pitch over time the state variables are:

$$(X_1, X_2, X_3) = [6.3889570e+06,\ 2.2584399e+04,\ -4.3213012e+03],\ m$$

$$(V_{X1}, V_{X2}, V_{X3}) = [5.0472863e+02,\ 3.7638634e+02,\ -3.4838204e+02]\ (\text{m/s})$$

corresponding to the data in Table 2.3:

2.3 The position/velocity variables in LBI frame of Problem 2.2 are transformed into ECI variables by the formulas (2.46)–(2.51) with the absolute longitude of the launch site λ_a and latitude L_0 as in Problem 1.2. The ECI coordinates are

$$(X, Y, Z) = [2.0926577e+06,\ 6.0274253e+06,\ -3.3200627e+05],\ m$$

Table 2.3 Data at pitch over

time (s)	Event	Altitude (km)	Rel. Vel (m/s)	Rel. fpa (deg)
21	End Pitch rotation	3.9	351.0	65.8
39	Pitch over	11.0	719.5	45.4

$$(V_X, V_Y, V_Z) = [-1.9403752e+02,\ 5.8344015e+02,\ -3.7381111e+02],\ \mathrm{m/s}$$

These are transformed into state variables, see (2.46)-(2.53):

$$(r, \lambda_a, L) = [6.3889984e+06\,m,\ 1.236e+00\,\mathrm{rad},\ -5.198e-02\,\mathrm{rad}]$$

$$(V_R, \gamma_R, \zeta_R) = [719.5\,\mathrm{m/s},\ 7.924e-01\,\mathrm{rad},\ -7.489e-01\,\mathrm{rad}]$$

These data are the initial condition of the numerical integration of the gravity turn equations (2.33), (2.45) with the remaining propellent of the first stage and the second and third stage boosts, without coasting between them.

The propellant remaining after 39 s is equal to $m_{p_1} = 5645$ kg, then the total mass at beginning of the gravity turn is $M_1 = 14067$ kg.

The second stage thrust (N) and propellant mass variation (kg/s) are interpolated as it follows:

$$T = q_5\, t^5 + q_4\, t^4 + q_3\, t^3 + q_2\, t^2 + q_1\, t + q_0\,,\ \ t \in [0, t_{b2}]\ \ t_{b_2} = 38.8\, s$$
$$q_5 = -0.1075,\ q_4 = 8.7624,\ q_3 = -254.0326,\ q_2 = 2898.0486,\ q_1 = -2795.3941,\ q_0 = 179471$$
$$m = m_{p20} - \dot{m}_2\, t,\quad m_{p20} = 3729\, kg,\ \ \dot{m}_2 = m_{p20}/t_{b_2} = 96.11\, kg/s$$

The initial mass after the first stage separation and ignition of the second stage is $M_2 = 6392$ kg

After the second stage burn-out and separation, the heat shields are also released, so the mass at the third stage ignition is $M_3 = 1848$ kg.

The third stage thrust (N) and propellant mass variation (kg/s) are interpolated as it follows:

$$T = q_{26}\, t^6 + q_{25}\, t.^5 + q24 * t.^4 + q23 * t.^3 + q22 * t.^2 + q21 * t + q20\,,\ \ t \in [0, t_{b_3}],\ t_{b_3} = 47.6\, s$$
$$q_{26} = -0.0014272,\ q_{25} = 0.1922,\ q_{24} = -9.8601,\ q_{23} = 242.23,\ q_{22} = -3000.7,\ q_{21} = 18765,\ q_{20} = 34325$$
$$m_{p_3} = m_{p30} - \dot{m}_3\, t,\ \ m_{p30} = 1286\, kg\ ,\ \ \dot{m}_3 = m_{p30}/t_{b_3} = 27.02\, kg/s$$

The following table shows state variables at relevant events (Table 2.4).

2.4 The variation of velocity generated by the engines of each of the three stages are

$$\Delta V_1 = V_{1f}\ ,\quad \Delta V_2 = V_{2f} - V_{1f}\ ,\quad \Delta V_3 = V_{3f} - V_{2f}$$

where V_{kf} are the velocities computed at the burn out of each stage.

Table 2.4 Data at burn out of the first three stages

Time (s)	Event	Altitude (km)	Rel. Vel (m/s)	Rel. fpa (deg)
80	First stage burn out	43.6	1548	31
119	Secont stage burn out	86.5	3796	24
166	Third stage burn out	176.5	6615	21

The analytic formula for the velocity of each stage in gravity turn is:

$$V_1 = A_1\, z_1^{\bar{n}_1}(1+z_1^2)\ ,\quad z_1 \in [0, z_{1f} = z_{20}]$$

$$V_2 = A_2\, z_2^{\bar{n}_2}(1+z_2^2)\ ,\quad z_2 \in [z_{20}, z_{2f} = z_{30}]$$

$$V_3 = A_3\, z_3^{\bar{n}_3}(1+z_3^2)\ ,\quad z_3 \in [z_{30}, z_{3f} = 1]$$

Then

$$V_{1f} = A_1 z_{20}^{\bar{n}_1}(1+z_{20}^2)\ ,\ V_{2f} = A_2 z_{30}^{\bar{n}_2}(1+z_{30}^2)\ ,\ V_{3f} = A_3 z_{3f}^{\bar{n}_3}(1+z_{3f}^2) = 2A_3 \tag{2.90}$$

In (2.90) the thrust parameters $\bar{n}_1, \bar{n}_2, \bar{n}_3$ are in Tables 1.5 and 1.8 and the unknowns $A_1, A_2, A_3, z_{20}, z_{30}$ are obtained solving the 5 Eqs. (2.71)–(2.75). Guess values of the velocities A_k are the Tziolkowki velocities V_{Tz_k} of Tables 1.5 and 1.8. The guess for z_{20}, z_{30} must be taken in the interval (0, 1), here the choice is: $z_{20} = 0.5$, $z_{30} = 0.5$.

For the Scout G launcher, convergence is achieved at the values

$$A_1 = 1761\,\text{m/s}\ ,\ A_2 = 3547\,\text{m/s}\ ,\ A_3 = 3846\,\text{m/s},\quad z_{20} = 0.8;\ z_{30} = 0.9$$

Note that the values z_{20}, z_{30} correspond to the flight path angles at the first and second burn out: $\gamma_{1f} = 11.4$ deg, $\gamma_{2f} = 6$ deg.

Formula (2.90) gives

$$V_{1f} = 1933\,\text{m/s}\ ,\ V_{2f} = 4287\,\text{m/s}\ ,\ V_{3f} = 7692\,\text{m/s}$$

and the variation of velocity generated by each stage is equal to:

$$\Delta V_1 = 1933\,\text{m/s}\ ,\ \Delta V_2 = 2354\,\text{m/s},\ \Delta V_3 = 3405\,\text{m/s}$$

The gravity losses of each stage are:

$$\Delta V_{g_1} = V_{Tz_1} - V_{1f} = 398\,\text{m/s}\ ,\ \Delta V_{g_2} = V_{Tz_2} - \Delta V_2 = 51\,\text{m/s}\ ,\ \Delta V_{g_3} = V_{Tz_3} - \Delta V_3 = 17\,\text{m/s}$$

Summing the gravity losses, one has:

$$\Delta V_g = \Sigma_{k=1}^{3} \Delta V_{Tz_k} - \Sigma_{k=1}^{3} \Delta V_k = \Sigma_{k=1}^{3} \Delta V_{Tz_k} - V_{3f} = 466\,\mathrm{m/s}$$

For the Vega launcher one obtains:

$$A_1 = 2968\,\mathrm{m/s}\ ,\ A_2 = 3019\,\mathrm{m/s}\ ,\ A_3 = 3812\,\mathrm{m/s}\ ,\ z_{20} = 0.6\ ,\ z_{30} = 0.8$$

Note that the values z_{20}, z_{30} correspond to the flight path angles at the first and second stage burn out: $\gamma_{20} = 25.14$ deg, $\gamma_{30} = 11.4$ deg.

The velocities at the end of each stage are:

$$V_{1f} = 2033\,\mathrm{m/s}\ ,\ V_{2f} = 4220\,\mathrm{m/s}\ ,\ V_{3f} = 7624\,\mathrm{m/s}$$

and the gravity losses are:

$$\Delta V_{g_1} = 752\,\mathrm{m/s}\,,\ \Delta V_{g_2} = 227\,\mathrm{m/s}\,,\ \Delta V_{g_3} = 103\,\mathrm{m/s}$$

Note that the value ΔV_{g_1} is close to the gravity loss of Table 2.1 that was obtained numerically evaluating the gravity loss along the nominal (polar) mission.

2.5 Flight path angles γ_{ex} at the third stage burn out are considered in the range $[0, 40°]$. The corresponding values $z_{3_f} = tan(\frac{\frac{\pi}{2}-\gamma_{ex}}{2})$ are inserted into the Eqs. (2.71)–(2.75). These are solved by the single guess $A_1 = \Delta V_{Tz_1}$, $A_2 = \Delta V_{Tz_2}$, $A_3 = \Delta V_{Tz_3}$ and $z_{20} = z_{30} = 0.5$. The solutions determine the velocity $V_{3_f} = A_3\, z_{3_f}(1 + z_{3_f}^2)$ and altitude h_{3_f} obtained by Eq. (2.62). Velocity and altitude at third stage burn out are plotted in Fig. 2.12.

2.6 The trajectories starting from the San Marco launch base are determined using a kick angle of 0.5 deg, a vertical arc duration of $t_v = 6\,\mathrm{s}$, an azimuth angle $\psi_A = 134$ deg and a pitch angular rate $\dot{\theta}_p = 3$ deg/s. Different durations of the pitch manoeuvre t_p are considered. These data are used to determine the pitch over time and the subsequent gravity turn arcs produced by the first, second and third stage with no coasting between them. This is done as in Problems 2.1, 2.2 and 2.3. At the end of the third stage burn all the state variables are calculated. The flight path angle γ_{ex} at the end of the third stage is shown in Fig. 2.13 for t_p ranging from 9 to 23 s.

The associated altitudes and the (relative) velocities are shown in Fig. 2.14.

References

1. Cornelisse, J.W., Shoyer, F.R., Wakker, K.F.: Rocket Propulsion and Spaceflight Dynamics. Pitman, Boston (1979)
2. Noton, M.: Spacecraft. Navigation and Guidance. Springer, Berlin (1988)
3. Vihn, N.X.: Optimal Trajectories in Atmospheric Flight. Elsevier, London (1981)
4. Calise, A., Melaned, N., Lee, S.: Design and evaluation of a three dimensional optimal ascent algorithm. J. Guidance, Control Dyn. 21(6) (1998)

5. Gath, C.A.: Optimization of launch vehicle trajectories with path constraints and coasting arcsJ. Guidance, Control Dyn. **24**(2) (2001)
6. Liu, P., Sun, H., Tsai, B.: Closed loop endoatmospheric ascent guidance. J. Guidance, Control Dyn. **26** (2003)
7. Liu,P., Pan, B.: Highly constrained optimal ascent guidance. J. Guidance, Control Dyn. **33** (2010)
8. Roh, W., Kim, Y.: Trajectory optimization of multistage launch vehicle using finite element and direct colocation methods. Eng. Optim. 34 (2002)
9. Jamilnia, R., Nagash, A.: Simultaneous optimization of staging and trajectory of launch vehicle using two different approaches. Aerospace Sci. Technol. **34** (2012)
10. Pontani, M.: Particle swarm optimization of ascent trajectories of multistage launch vehicles. Acta Astronaut. **94**, 852–864 (2014)
11. Pallone, M., Pontani, M. Teofilatto, P.: Performance evaution methodology for multistage launch vehicles with high fidelity modeling. Acta Astronaut. **151** (2018)
12. Culler, G., Fried, B.: Universal gravity turn trajectories. J. Appl. Phys. **28**, 672–676 (1957)
13. Di Sotto, E, Teofilatto, P.: Semi-analytic formulas for launcher performance evaluation. J. Guidance, Control Dyn. **25** (2002)
14. Teofilatto, P.: A paper and pencil method for evaluating launcher performances. In: Proceedings of the International Astronautical Conference, Adelaide, Australia 25–29 September 2017 (2017)
15. Teofilatto, P., Carletta, S., Pontani, M.: Analytic derivation of ascent trajectories and performances of launch vehicles. Appl. Sci. **12** (2022)

Chapter 3
Ascent Trajectories: Exo-Atmospheric Arc

Abstract The trajectory of the launcher in vacuum space can be determined by solving an optimal control problem. The formulation of the problem leads to the Primer vector theory, where the optimal thrust direction, the number and duration of coasting arcs are determined by the appropriate initial conditions of the Lagrangian multiplier corresponding to the velocity vector. The numerical methods required to solve the optimal problem, and hence to derive the nominal optimal trajectory, are not applicable to the flight of a space launcher. Simplified and direct guidance algorithms are required, which are suitable for the implementation of real-time guidance algorithms on the on-board computer. In this Chapter, simple criteria for the duration of the coasting arc after the atmospheric flight are given and the final injection into orbit is studied by means of a time-optimal guidance implemented with simplified models of the launcher dynamics. Furthermore, the Q-guidance algorithm is presented: this guidance algorithm was introduced in the Apollo missions and can be implemented in real time by analytical formulae. Some current launchers implement a real-time tangent linear law for the trust guidance to injection into orbit. This guidance is described here. Applications are given in the problem posed and solved at the end of the chapter.

3.1 Introduction

In the exo-atmospheric arc of the trajectory the guidance system is free to direct the launcher, hence the thrust, in the most convenient direction to achieve orbit insertion. In general, most convenient means maximising the payload mass that can be placed into orbit. This is a classic problem of space trajectory optimisation, a long-studied subject, see for example [1–3]. The solution method is based on a generalisation of the calculus of variations (Primer vector theory) and introduces additional functions called Lagrangian multipliers. In this formalism, the control functions, i.e. the thrust intensity law and the thrust direction in time, are replaced by the Lagrangian multipliers which are solutions of known differential equations. The problem of finding an optimal trajectory in the (infinite dimensional) space of the control functions is translated into the problem of finding the initial conditions of the Lagrangian multi-

P. Teofilatto, *Trajectories of Launch and Reentry Systems*, Springer Aerospace Technology, https://doi.org/10.1007/978-3-031-75760-0_3

pliers. Thus, the calculus of variations greatly simplifies the optimisation problem, reducing the search for the solution from an infinite dimensional space to a finite dimensional space. However, the search for the unknown initial conditions of the Lagrangian multipliers is not easy and relies on numerical iterative methods whose convergence can be difficult and time consuming. As such it can not be implemented on the flight program of a launcher.

A possible strategy is then to compute an optimal nominal trajectory before the flight, load its parameters into the on-board computer memory and implement a control algorithm to correct the current state, as measured by the navigation algorithm, to the nominal state within the duration of the time control step.

Such guidance algorithms are called perturbative because the objective is to compensate for the perturbations that may occur with respect to the nominal trajectory.

On the other hand, adaptive guidance algorithms generate a new trajectory from the current state to the target state. Of course, "sub-optimal" trajectories are generated simplifying the dynamic model. These trajectories are described by explicit formulae, so that real-time guidance can be implemented on board and updated at appropriate time intervals to ensure convergence of the trajectory to the target orbit. Since the Apollo missions, adaptive guidance algorithms have been implemented, some of which are described in this chapter.

Before implementing adaptive guidance, coasting manoeuvre is performed to gain altitude at expense of reducing speed. If the atmospheric phase of the trajectory has been well designed, the coasting time should be such that the target orbital altitude is reached by the coasting arc and the guidance of the engine of the last stage is able to gain orbital velocity. In general, a trade-off between altitude gain and velocity loss is required in the selection of the coasting duration.

After coasting phase analysis, three types of adaptive guidance are described in this Chapter: (i) a simplified time optimal guidance, (ii) the Q-guidance, (iii) the iterative tangent linear guidance.

3.2 General Equation of Fight in Vacuum Space

The equations of motion considered so far have been defined with respect to the relative to the wind velocity and in the moving frame RVF.

In the exo-atmospheric phase the aerodynamics is negligible and there is no reason to consider the relative velocity. The relationships between the relative velocity V_R, the associated co-azimuth ζ_R, flight path angle γ_R and the corresponding absolute values (V, ζ, γ) are as follows:

$$V = \sqrt{V_R^2 + (\omega_E\, r \cos L)^2 + 2V\, r\omega_E\, \cos L\, \cos\gamma_R \cos\zeta_R}$$

$$\gamma = arcsin[\frac{V_R \sin\gamma_R}{V}]$$

$$\cos\zeta = \frac{V_R \cos\gamma_R \cos\zeta_R + \omega_E\, r\cos L}{V\cos\gamma}, \quad \sin\zeta = \frac{V_R\cos\gamma_R \sin\zeta_R}{V\cos\gamma}$$

The equation of motion in absolute state variables $(r, \lambda_a, L, V, \gamma, \zeta)$ can be derived as in Sect. 2.3, see for instance [4]. The final result can also obtained setting in Eqs. (2.35), (2.36), (2.37) the Earth angular velocity ω_E to zero and neglecting aerodynamics in (2.42), (2.43), (2.44). Note that in these hypotheses the relative and absolute velocities coincide.

The dynamics is then represented by the equations:

$$\begin{cases} \dot V = -\frac{\mu}{r^2}\sin\gamma + \frac{T}{m}\cos\alpha_T\cos\beta_T \\ \dot\gamma = \frac{V}{r}\cos\gamma - \frac{\mu}{V r^2}\cos\gamma + \frac{T}{V m}\sin\alpha_T\cos\beta_T \\ \dot\zeta = -\frac{V}{r}\cos\gamma\tan L\cos\zeta + \frac{T}{V\cos\gamma\, m}\sin\beta_T \end{cases} \tag{3.1}$$

to be integrated with the kinematic equations

$$\begin{cases} \dot r = V\sin\gamma \\ \dot\lambda_a = \omega_E + \frac{V\cos\gamma\cos\zeta}{r\cos L} \\ \dot L = \frac{V\cos\gamma\sin\zeta}{r} \end{cases} \tag{3.2}$$

and some choice of the guidance angles α_T, β_T and thrust intensity T.

Note, however, that the final target, that is the orbit to be achieved, is represented in the Earth Centered Inertial (ECI) frame. The coordinates of the position in the ECI frame are related to the state coordinates by the relationships:

$$X = r\cos L\cos\lambda_a$$

$$Y = r\cos L\sin\lambda_a$$

$$Z = r\sin L$$

The velocity vector in ECI frame is related to the state coordinates by the rotations:

$$\mathbf{V} = [V_X\ V_Y\ V_Z]\begin{bmatrix}\hat c_1\\ \hat c_2\\ \hat c_3\end{bmatrix} = [V\cos\gamma\cos\zeta\quad V\cos\gamma\sin\zeta\quad V\sin\gamma]\begin{bmatrix}\hat r\\ \hat E\\ \hat N\end{bmatrix}$$

By inversion of the rotations from ECI to the LHF:

$$\begin{bmatrix}\hat c_1\\ \hat c_2\\ \hat c_3\end{bmatrix} = R_L^2\ R_{\lambda_a}^3\begin{bmatrix}\hat r\\ \hat E\\ \hat N\end{bmatrix}$$

the ECI velocity coordinates are related to the state velocity coordinates by:

$$[V_X\ V_Y\ V_Z] = [V\cos\gamma\cos\zeta\quad V\cos\gamma\sin\zeta\quad V\sin\gamma]\,(R^3_{\lambda_a})^t\,(R^2_L)^t$$

To compare the relative and absolute coordinates, consider the trajectory of the Scout launcher from Problem 2.3 (payload mass equal to 150 kg, including payload adapter). Figure 3.1 compares the relative and absolute velocity. The relative velocity starts from zero, while the absolute velocity is initially equal to the velocity of rotation of the Earth at the San Marco launch site. As the trajectory progresses, the two velocities converge as the acceleration is the same. Figure 3.2 compares the relative and absolute flight path angles. The relative flight path angle starts at 90 deg, which is the direction of the relative velocity during the vertical arc. The direction of the absolute velocity is initially towards East, so the absolute flight path angle is initially zero.

The kinetic and dynamic equations in ECI frame are:

$$\begin{cases} \frac{dX}{dt} = V_X \\ \frac{dY}{dt} = V_Y \\ \frac{dZ}{dt} = V_Z \\ \frac{dV_X}{dt} = -\frac{\mu}{r^3}X + \frac{T}{m}l_X \\ \frac{dV_Y}{dt} = -\frac{\mu}{r^3}Y + \frac{T}{m}l_Y \\ \frac{dV_Z}{dt} = -\frac{\mu}{r^3}Z + \frac{T}{m}l_Z \end{cases} \tag{3.3}$$

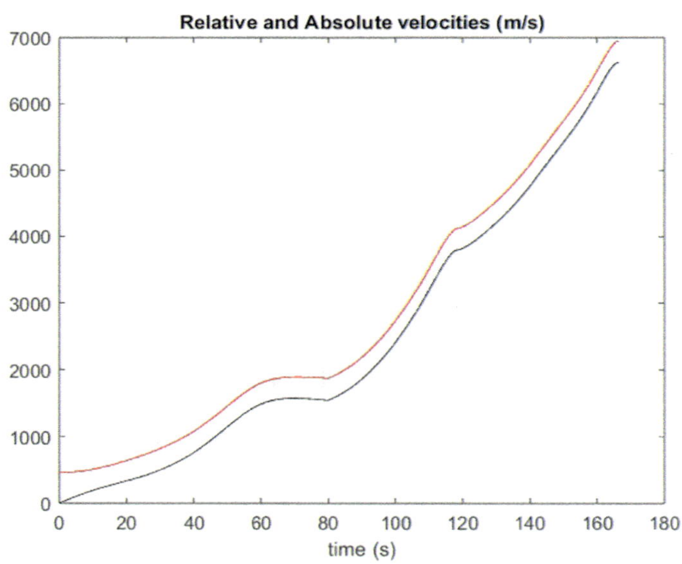

Fig. 3.1 The relative and absolute velocity of the Scout G trajectory with conditions defined in Problem 2.3

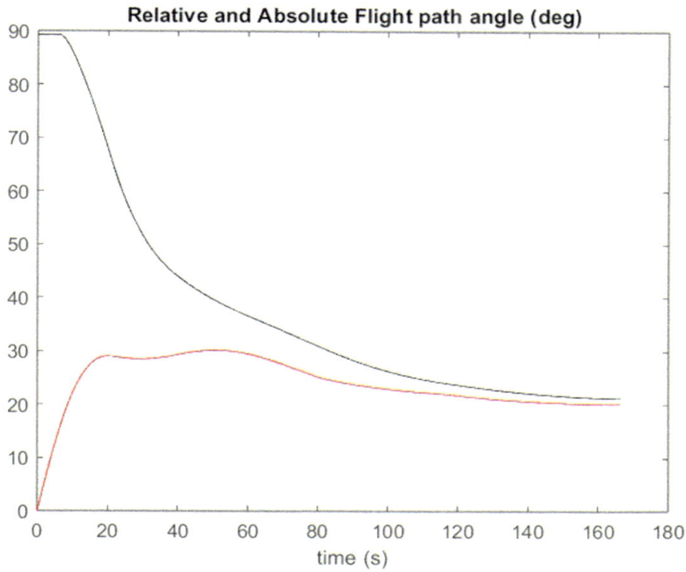

Fig. 3.2 The relative and absolute flight path angle of the Scout G trajectory with conditions defined in Problem 2.6

where (l_X, l_Y, l_Z) determine the direction of the thrust in ECI frame, $\sqrt{l_x^2 + l_y^2 + l_z^2} = 1$.

These variables, together with the thrust intensity, represent the control variables to determine an optimal trajectory from the initial exo-atmospheric state

$$\mathbf{r}(t_0) = \mathbf{r}_0 \quad , \quad \mathbf{V}(t_0) = \mathbf{V}_0 \tag{3.4}$$

to the final state

$$\mathbf{r}(t_f) = \mathbf{r}_f \quad , \quad \mathbf{V}(t_f) = \mathbf{V}_f \tag{3.5}$$

minimizing, for instance, the Tziolkowski variation of velocity produced by the engines:

$$J = \int_{t_0}^{t_f} \frac{T}{m} dt$$

To address the optimal control problem, the following variable is introduced [5]:

$$C(t) = \int_{t_0}^{t} \frac{T}{m} dt$$

and added to the state equations:

$$\begin{cases} \dfrac{d\mathbf{r}}{dt} = \mathbf{V} \\ \dfrac{d\mathbf{V}}{dt} = -\dfrac{\mu}{r^3}\mathbf{r} + \dfrac{\mathbf{T}}{m} \\ \dot{C} = \dfrac{T}{m} \end{cases} \tag{3.6}$$

so that the minimum of a final state variable is researched:

$$J = C_f$$

To determine optimal trajectories, a generalization of the calculus of variation method is applied, and the three-dimensional Lagrangian multipliers are introduced [5]:

$$\boldsymbol{\lambda}_{\mathbf{r}} = (\lambda_X, \lambda_Y, \lambda_Z)$$

$$\boldsymbol{\lambda}_{\mathbf{V}} = (\lambda_{V_X}, \lambda_{V_Y}, \lambda_{V_Z})$$

with the Hamilton function

$$H = \boldsymbol{\lambda}_{\mathbf{r}} \cdot \underline{\dot{r}} + \boldsymbol{\lambda}_{\mathbf{V}} \cdot \underline{\dot{V}} + \lambda_C \dot{C}$$

that is

$$H = \boldsymbol{\lambda}_{\mathbf{r}} \cdot \mathbf{V} + \boldsymbol{\lambda}_{\mathbf{V}} \cdot \left(-\frac{\mu}{r^3}\mathbf{r} + \frac{\mathbf{T}}{m}\right) + \lambda_C \frac{T}{m}$$

Let us split H in a term depending on control and a term, K, independent on the control; the control variables in this problem are the thrust modulus and direction:

$$H = \boldsymbol{\lambda}_{\mathbf{V}} \cdot \frac{\mathbf{T}}{m} + \lambda_C \frac{T}{m} + K$$

To find the minimum of H, consider the conditions to have the product $\boldsymbol{\lambda}_{\mathbf{V}} \cdot \frac{\mathbf{T}}{m}$ minimized: the minimum occurs when $\boldsymbol{\lambda}_{\mathbf{V}}$ is parallel and opposite with respect to $\mathbf{T}$: $\hat{\lambda_V} = -\hat{T}$. Then the lagrangian vector $\boldsymbol{\lambda}_{\mathbf{V}}$ determines the direction of the optimal thrust and for this reason it is called "primer vector" by the researcher D. Lawden who developed this theory. With this optimal control the Hamilton function becomes:

$$H^* = (\lambda_C - \lambda_V)\frac{T}{m} + K$$

and this is minimized according to the following rules:

1. If $(\lambda_C - \lambda_V) < 0 \Rightarrow T^* = T_{max}$ maximum thrust arch
2. If $(\lambda_C - \lambda_V) > 0 \Rightarrow T^* = 0$ coasting arch
3. If $(\lambda_C - \lambda_V) = 0 \Rightarrow T^* = ?$ singular arch

It follows that the knowledge of the lagrangian multipliers determines the intensity T^* and direction $-\hat{\boldsymbol{\lambda}}_{\mathbf{V}}$ of the optimal thrust. The multipliers follow the equations:

$$\dot{\underline{\boldsymbol{\lambda}}}_{\mathbf{r}} = -\frac{\partial H}{\partial \mathbf{r}} = \frac{\partial}{\partial \mathbf{r}}(\frac{\mu}{r^3}\mathbf{r}) \cdot \boldsymbol{\lambda}_{\mathbf{V}}$$

$$\dot{\underline{\boldsymbol{\lambda}}}_{\mathbf{V}} = -\frac{\partial H}{\partial \mathbf{V}} = -\boldsymbol{\lambda}_{\mathbf{r}}$$

$$\dot{\lambda}_C = -\frac{\partial H}{\partial C} = -\frac{\partial H}{\partial m}\frac{\partial m}{\partial C}$$

The first derivative is computed as the gravity gradient and it is equal to the tensor:

$$\frac{\partial}{\partial \mathbf{r}}(\frac{\mu}{r^3}\mathbf{r}) = -\frac{\mu}{r^3}(3\hat{r}\hat{r} - \mathbf{u})$$

with $\mathbf{u}$ representing the identity matrix in the matrix representation of the gravity gradient tensor.

To compute the derivative of λ_C, note that:

$$\frac{\partial H}{\partial m} = -\frac{T}{m^2}(\lambda_C - \lambda_V)$$

Moreover, if $T = -g\, Isp\, \dot{m}$ the variable C is equal to $C = g\, Isp\, log(\frac{m_0}{m(t)})$ then

$$\frac{\partial m}{\partial C} = \left(\frac{\partial C}{\partial m}\right)^{-1} = \left(-\frac{g\, Isp}{m}\right)^{-1}$$

Therefore the equations of the lagrangian multipliers are:

$$\begin{cases} \dot{\underline{\boldsymbol{\lambda}}}_{\mathbf{r}} = \frac{\mu}{r^3}(3\hat{r}\hat{r} - \mathbf{u})\boldsymbol{\lambda}_V \\ \dot{\underline{\boldsymbol{\lambda}}}_{\mathbf{V}} = -\boldsymbol{\lambda}_{\mathbf{r}} \\ \dot{\lambda}_C = -(\lambda_C - \lambda_V)\frac{T}{m}\frac{1}{g\, Isp} \end{cases} \tag{3.7}$$

where the matrix $\frac{\mu}{r^3}(3\hat{r}\hat{r} - \mathbf{u})$ is the gravity gradient matrix and (3.7) can be written as

$$\begin{cases} \ddot{\underline{\lambda}}_{\mathbf{V}} = -\frac{\mu}{r^3}(3\hat{r}\hat{r} - \mathbf{u})\mathbf{r} = -\frac{\mu}{r^3}[(3\hat{r}\,\boldsymbol{\lambda}_V)\hat{r} - \boldsymbol{\lambda}_V] \\ \dot{\lambda}_C = -(\lambda_C - \lambda_V)\frac{T}{m}\frac{1}{g\,Isp} \end{cases} \tag{3.8}$$

Equations (3.6) and (3.8) are integrated with the state initial conditions (3.4) and initial conditions of the primer vector:

$$\boldsymbol{\lambda}_V(t_0) \quad , \quad \dot{\boldsymbol{\lambda}}_V(t_0)$$

so that the final conditions (3.5) are met and minimum of C_f is achieved.

The optimal thrust direction is equal to the direction $\hat{\lambda}_V(t)$ obtained integrating (3.6) and (3.8) and the thrust intensity is obtained according to maximum/coasting/singular arc conditions found according to the above rules 1–3.

However the initial conditions of the primer vector to achieve the final conditions (3.5) are unknown and the determination of the optimal trajectory depends on the correct definition of such initial conditions, a two point boundary problem difficult to solve.

Some help can be provided by the knowledge of the final values of some lagrangian multiplier: if the final state x_{k_f} is not specified, the value of the corresponding lagrangian multiplier is:

$$\lambda_{x_{k_f}} = -\frac{\partial J}{\partial x_{k_f}} \tag{3.9}$$

In the present case, the equations of the multipliers allow to make some general considerations and define the structure of the optimal control.

Namely, let at initial time have $(\lambda_C - \lambda_V) < 0$, so one has maximum thrust as the starting of the guidance, see rule 1. Then: $\lambda_V > \lambda_C$ and λ_C is an increasing function of the time, as can be seen in (3.7). It may be a time t_1 so that $\lambda_C > \lambda_V$ corresponding to coasting arc. In this case the thrust intensity is zero (see rule 2) and the Eq. (3.7) states $\dot{\lambda}_C = 0$, so λ_C is constant. Then the behaviour of the optimal multipliers can be as shown in Fig. 3.3 where optimal coasting times separate arcs at maximum thrust. The switching times t_1, t_2, t_3 from maximum thrust to coasting correspond to the physical unknown variables of the optimal problem.

If optimality is sought for a trajectory of short range, for instance the arc of the last stage boost just before the injection into orbit, the approximation with the flat Earth approximation may be acceptable and the Hamilton function H does not depend on $\mathbf{r}$. Then $\dot{\lambda}_{\mathbf{r}} = 0$ and the lagrangian multipliers are:

$$\boldsymbol{\lambda}_{\mathbf{r}}(t) = \boldsymbol{\lambda}_{\mathbf{r}_0}$$

$$\boldsymbol{\lambda}_{\mathbf{V}}(t) = -\boldsymbol{\lambda}_{\mathbf{r}_0} t + \boldsymbol{\lambda}_{\mathbf{V}_0}$$

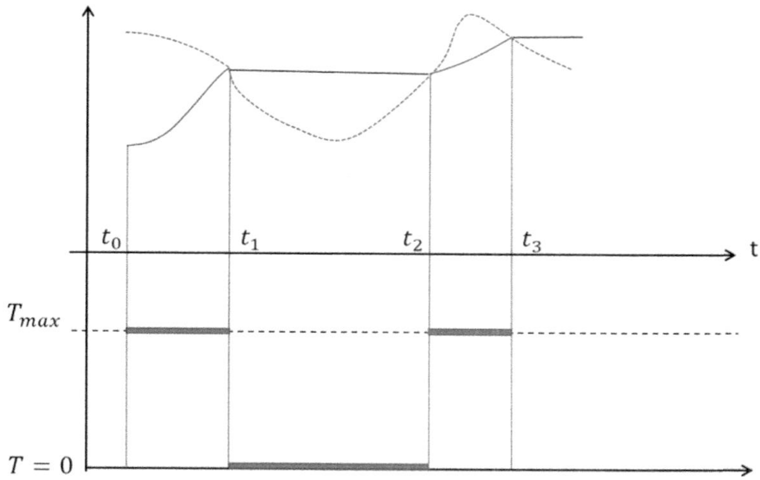

Fig. 3.3 The time history of the Primer vector λ_V (dotted) and the Lagrangian multiplier λ_C and the related optimal thrusting

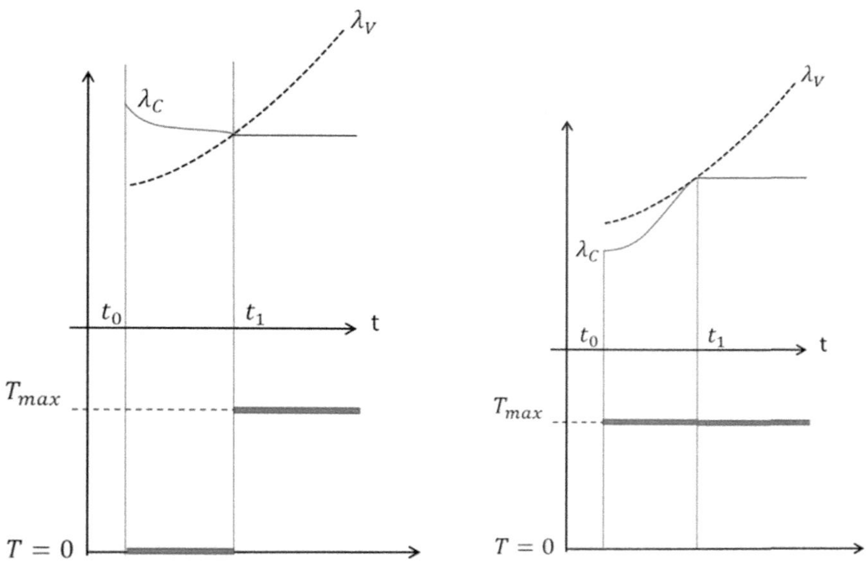

Fig. 3.4 The optimal thrusting strategy for flat Earth: coast-burn (left) or single burn (right)

It follows that the primer vector modulus λ_V is a monotonic function of time. As a result only one optimal burning strategy is possible: a single coasting followed by a single burn if $\lambda_{C_0} > \lambda_{V_0}$, a single burn only if $\lambda_{C_0} < \lambda_{V_0}$, see Fig. 3.4.

3.3 Coasting Arcs

At the end of the atmospheric arc the rocket has gained kinetic energy but its altitude is generally low. See for instance the Problem 2.3 referred to the Scout launcher after the thrust of the first three stages. Note the low altitude achieved: this can not be increased to the final orbit altitude only by the fourth stage burn.

It is then useful to consider a coasting arc to gain altitude without any thrust at the cost of a certain reduction in speed.

For instance, if a coasting arc of 400 s is added to the trajectory of Figs. 3.1 and 3.2, the following Figs. 3.5 and 3.6 show how the altitude increases as the velocity decreases. The variation of the flight path angle is in Fig. 3.7. Figure 3.8 shows the ground track of the launcher trajectory until the coasting phase.

The duration of the coasting arc is a parameter to be optimised and depends on the initial state (3.4) and on the target orbit: it can be derived solving the optimisation problem of Sect. 3.2.

A simpler approach is to perform a parametric search on the coasting duration t_c and to implement the guidance of the last stage to achieve the required orbit. According to different coasting times the injection into the target orbit may be either not feasible or achieved with different propellant consumption, and the best solution can be selected.

In fact the trajectory in the coasting phase is an arc of Kepler orbit determined by the initial exo-atmospheric conditions at burn out of the stage: $\mathbf{r}_{ex} = (X_{ex}, Y_{ex}, Z_{ex})$, $\mathbf{V}_{ex} = (VX_{ex}, VY_{ex}, VZ_{ex})$ (coordinates in ECI frame). The orbit is planar with

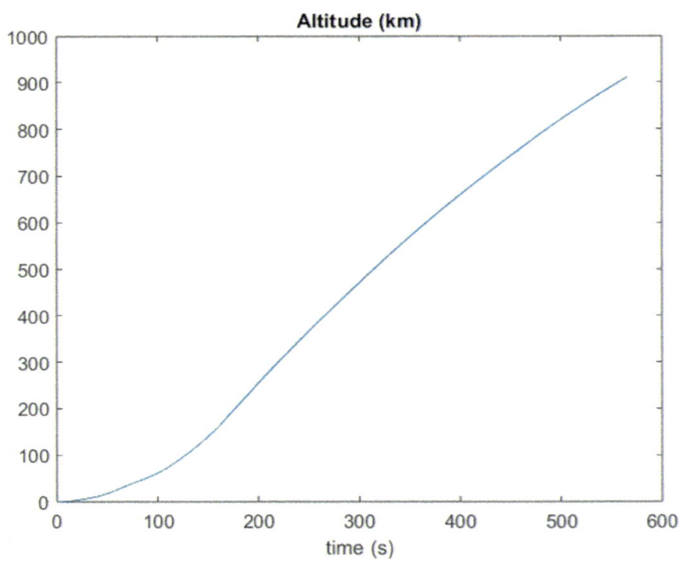

Fig. 3.5 Altitude (km) of the trajectory of Problem 2.3 including the coasting

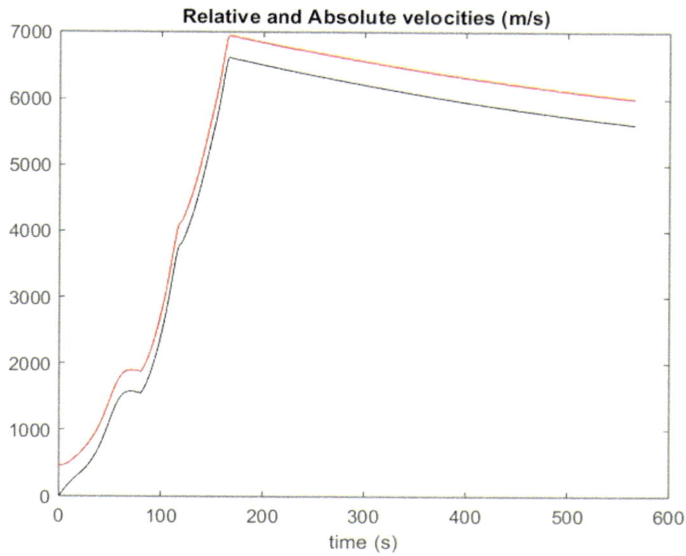

Fig. 3.6 Absolute and relative velocity (m/s) the trajectory of Problem 2.3 including the coasting

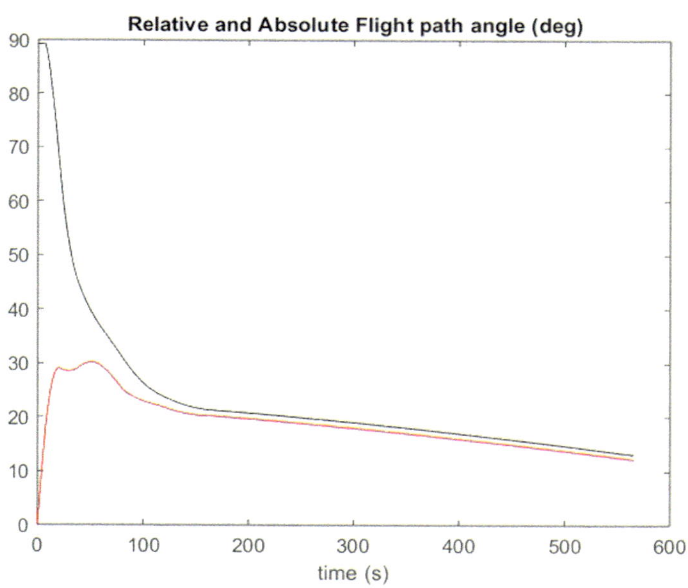

Fig. 3.7 Absolute and relative flight path angle (deg) of the gravity turn trajectory of Problem 2.3 including the coasting

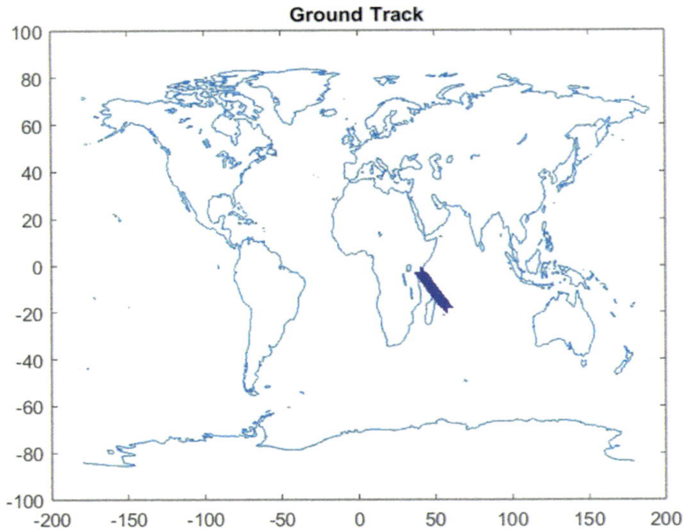

Fig. 3.8 Ground track of the Scout launcher trajectory of Problem 2.3 including the coasting

parameters that are defined by three integrals of the motion [6]: the energy E, the angular momentum vector $\mathbf{h}$ and the eccentricity vector $\mathbf{e}$. In particular E and $\mathbf{h}$ are defined:

$$E = \frac{1}{2}V^2 - \frac{\mu}{r}$$

$$\mathbf{h} = \mathbf{r} \wedge \mathbf{V}$$

From the energy it is possible to derive the semiaxis:

$$a = -\frac{\mu}{2E}$$

and from the semiaxis and the norm of $\mathbf{h}$ the eccentricity follows:

$$e = \sqrt{1 - \frac{h^2}{\mu\, a}}$$

Since E and h are constant during coasting, the semiaxis and the eccentricity are also constant during the entire coasting trajectory.

In fact all the parameters of the orbit corresponding to the state $(\mathbf{r}_{ex}, \mathbf{V}_{ex})$ are constant during the coasting arc. Figure 3.9 shows the semiaxis and eccentricity variation of the trajectory of Figs. 3.5, 3.6, the Fig. 3.10 shows the pericenter and the apocenter altitude and the Fig. 3.11 shows the RAAN and inclination angle. Note that

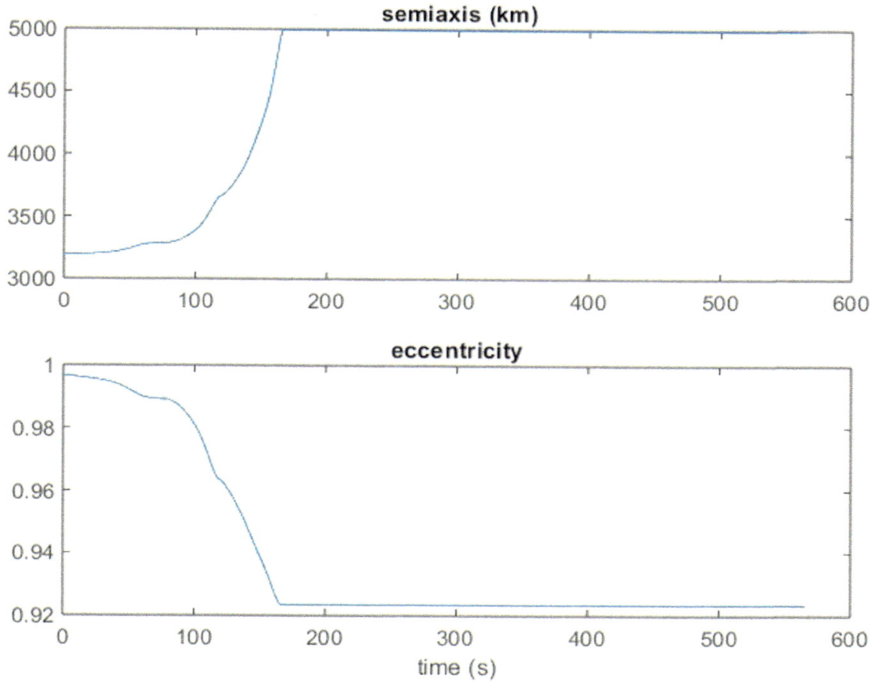

Fig. 3.9 Semiaxis (km) and eccentricity during the flight including coasting

during the coasting time the orbit parameters of Figs. 3.9, 3.10 and 3.11 are constant. The orbit inclination $i = 40°$ is achieved by the selected azimuth of launch according to the Equation of the Launch Base (see Problem 1.1) and the RAAN corresponds to the selected time of launch (see Problem 1.2, descending node, $t_0 = 0$).

Suppose that the target orbit is a circular orbit with inclination 40 deg, RAAN of 250 deg and altitude $h = 650\,\text{km}$. The out of plane orbit parameters are already achieved by the choice of launch time and azimuth angle, see Fig. 3.11. However, the in plane parameters semiaxis $a = 650 + R_E = 7028\,\text{km}$ and eccentricity $e = 0$ are not achieved, see Fig. 3.9. To obtain the correct parameters, different values of the flight path angle at exit of the atmospheric phase γ_{ex} and of the coasting time t_c must be chosen.

To have some hints, suppose that the last stage achieves the injection into the orbit by a gravity turn guidance and use (2.58), (2.59), (2.62) to determine, within the approximations of the analytical formulae, the state at ignition of the last stage.

Let $\bar{n}$ be the average thrust to weight ratio of the last stage: at the orbit injection we assume $z = 1$ and $V = V_C = \sqrt{\frac{\mu}{a}} = 7514\,\text{m/s}$. Then formula (2.58) gives

$$A = \frac{V_C}{2}$$

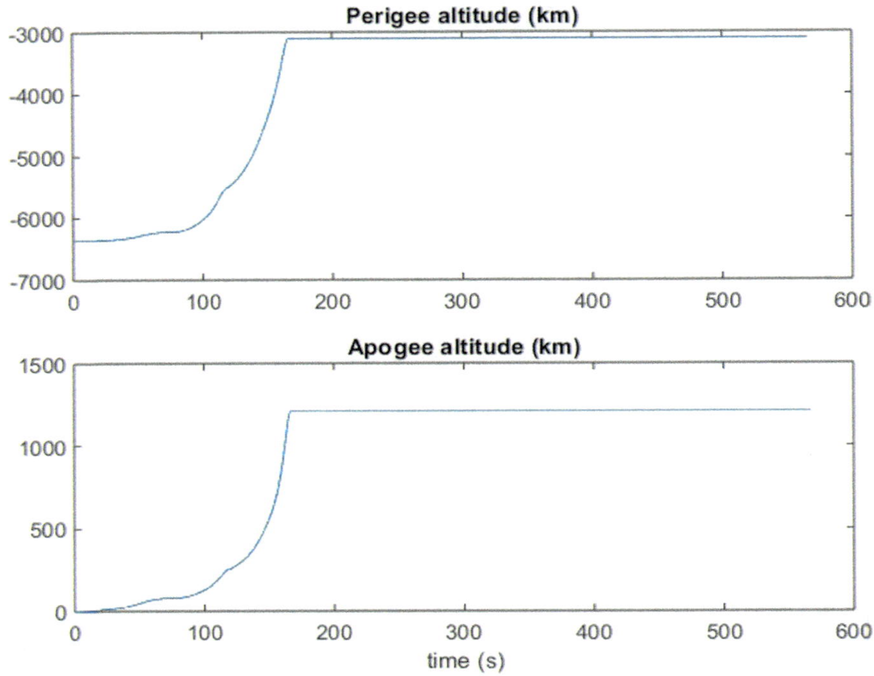

Fig. 3.10 Pericenter and apocenter altitudes (km) during the flight including coasting

and the velocity at the last stage ignition is:

$$V_0 = \frac{V_C}{2} z_0^{\bar{n}-1}(1 + z_0^2) \tag{3.10}$$

Formula (2.59) gives

$$t_b = \frac{A}{g}[\frac{z^{\bar{n}-1}}{\bar{n}-1} + \frac{z^{\bar{n}+1}}{\bar{n}+1}]_{z_0}^{z_f} \tag{3.11}$$

where $A = \frac{V_C}{2}$ and $z_f = 1$.

This is an equation for z_0 and its solution give the flight path angle γ_0 at the ignition of the last stage, that is at the end of the coasting phase. Formula (3.10) gives the velocity and by formula (2.62) one has the altitude at last boost:

$$h_0 = h - \frac{A^2}{g}[\frac{z^{2\bar{n}-2}}{2\bar{n}-2} + \frac{z^{2\bar{n}+2}}{2\bar{n}+2}]_{z_0}^{z_f} \tag{3.12}$$

where h is the target altitude.

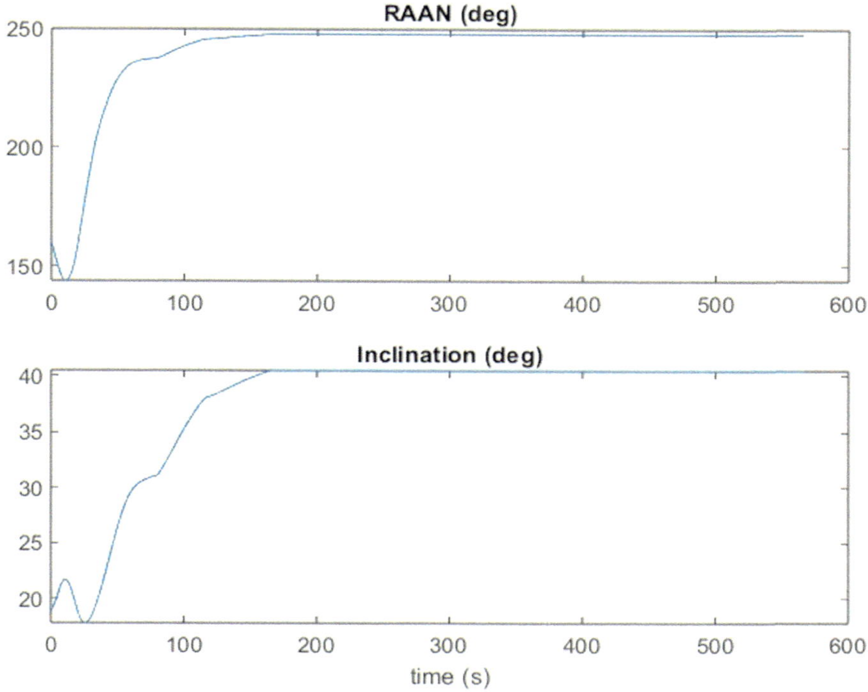

Fig. 3.11 RAAN and inclination (deg) during the flight including coasting

The values (V_0, h_0, γ_0) can be considered as estimated values at the end of the coasting for the injection into orbit by gravity turn performed by the last stage of the launcher. According to Problem 3.1 the state before the ignition of the Scout last stage are: $V_0 = 5607\,\text{m/s}$, $h_0 = 646$, $\gamma_0 = 2.33\,\text{deg}$. That is, the orbit injection is done by a sort of Hohmann transfer, with the required altitude reached by the coasting, and the velocity is gained by the last stage boost with almost no loss since the gravitational loss is small ($\gamma \sim 0$) and the trusting along the velocity direction produce no misalignment loss. Then the ΔV provided by the last stage is close to the Tziolkowski $\Delta V \sim 2000\,\text{m/s}$.

Coasting with maximum altitude equal to $h = 650\,\text{km}$ can be achieved by the Scout launcher with the time duration of the pitch manoeuvre $t_p = 17.8\,\text{s}$, corresponding to $\gamma_{ex} = 13.2\,\text{deg}$, and coasting time $t_c = 677s$.

Figures 3.12, 3.13, 3.14, 3.15, 3.16, show the altitude, velocity, flight path angle and orbit parameters of this trajectory. Figure 3.17 shows that the apogee has reached the target altitude h while a perigee increment to arrive to h must be performed by the last boost.

Table 3.1 provides state variables at relevant events.

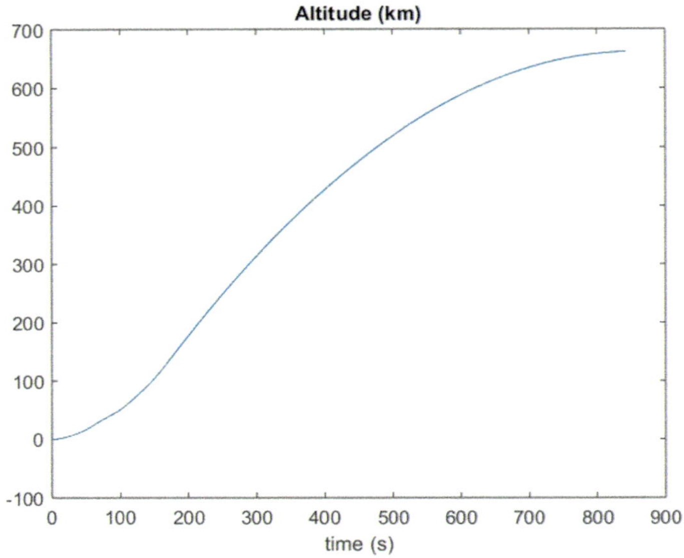

Fig. 3.12 Altitude (km) of the trajectory with $t_p = 17.8$ s (including the coasting of 677 s)

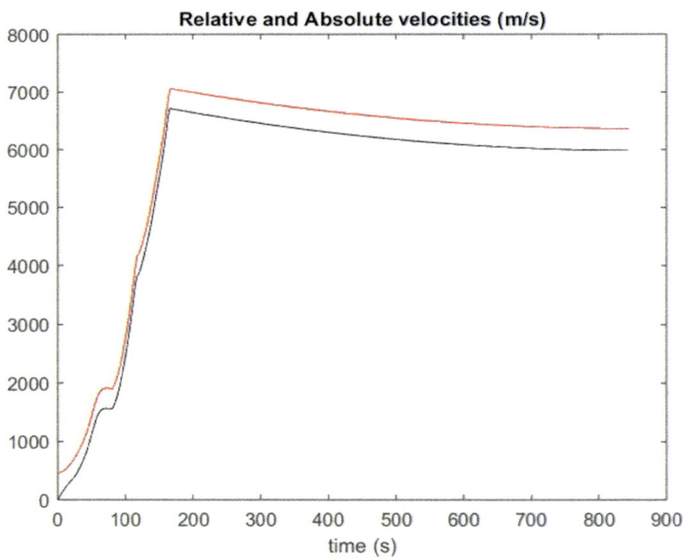

Fig. 3.13 Absolute and relative velocity (m/s) the trajectory with $t_p = 17.8$ s (including the coasting of 677 s)

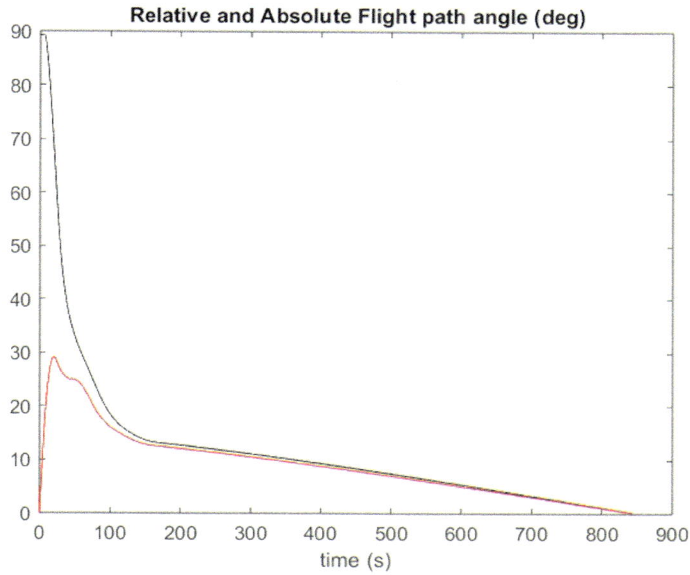

Fig. 3.14 Absolute and relative flight path angle (deg) of the trajectory with $t_p = 17.8$ s (including the coasting of 677 s)

3.4 Time Optimal Guidance to Orbit Injection

At the end of coasting the position and velocity of the launcher are in the target plane and close to the orbit injection. Namely most of the last stage boost is used to achieve the correct final velocity, since the altitude has been gained by the coasting arc.

For the orbit insertion is convenient to introduce the target plane determined by the required inclination i and RAAN Ω. The rotations R^3_Ω, R^1_i project the ECI coordinates onto the target plane: these will have zero out of plane position and and velocity, and only the planar coordinates (x_T, y_T, u_T, v_T) can be considered.

The position of the rocket (x_c, y_c) at the end of coasting in the target plane is determined by the angle ϕ taken from the nodal line:

$$\cos\phi = \frac{x_T}{r}$$

$$\sin\phi = \frac{y_T}{r}$$

and the rotation of the angle ϕ around the normal to the orbital plane $\hat{h}$ moves the nodal line direction to the radial direction $\hat{r}_c$ at the end of coasting, so determining the orbital frame $(\hat{r}_c, \hat{\theta}_c, \hat{h})$ which is called Target Plane Inertial (TPI) frame with coordinates (x, y, u, v) and dynamics given by:

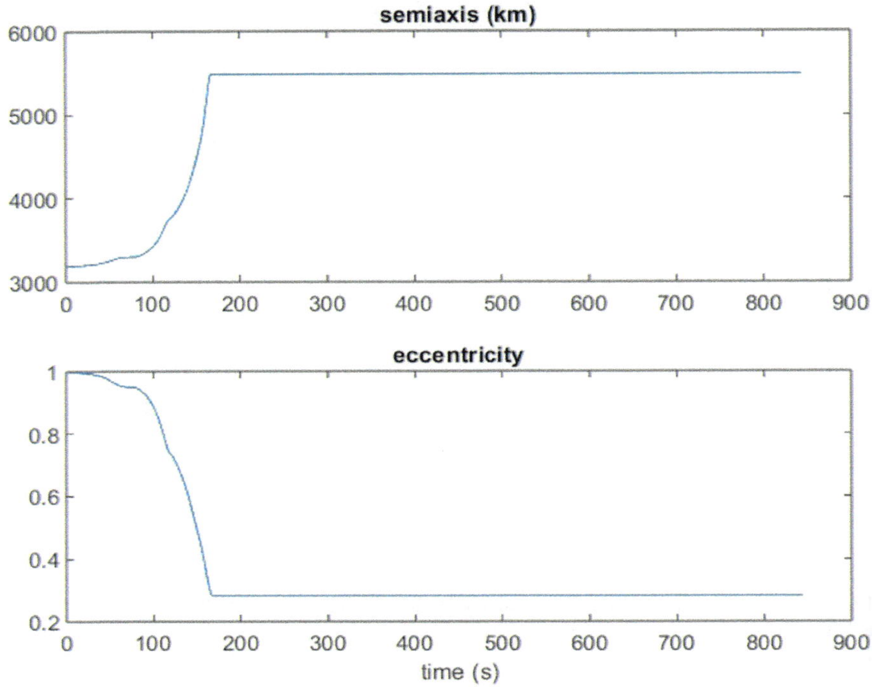

Fig. 3.15 Semiaxis (km) and eccentricity during the flight with $t_p = 17.8$ s (including coasting of 677 s)

$$
\begin{aligned}
\dot{x} &= v_x \\
\dot{y} &= v_y \\
\dot{v}_x &= g\bar{n}\sin\theta - g \\
\dot{v}_y &= g\bar{n}\cos\theta
\end{aligned}
\tag{3.13}
$$

with θ defining the thrust direction with respect to the horizontal direction $\hat{\theta}_c$, the gravity g is the average gravity from the position r_c to the final position r_f: $g = \frac{1}{2}(\frac{\mu}{r_c^2} + \frac{\mu}{r_f^2})$ and $\bar{n}$ is the average thrust over weight.

The injection into orbit is described by the following optimal problem: the initial state is given $\mathbf{X}_0 = (x_0, y_0, vx_0, vy_0)$ (the state after coasting). The final conditions are assigned $\mathbf{X}_f = (x_f = r_f, v_{xf} = 0, v_{yf} = \sqrt{\frac{\mu}{r_f}}))$ (the injection into circular orbit of radius r_f).

The task is to identify the guidance law $\theta(t)$ able to drive the last stage from $\mathbf{X}_0$ to $\mathbf{X}_f$ in minimum time. To solve the problem, the Hamilton function $H = \boldsymbol{\lambda}\dot{\mathbf{X}}$ is introduced [7]:

$$H = \lambda_x\, v_x + \lambda_y\, v_y + \lambda_{v_x}(g\bar{n}\sin\theta - g) + \lambda_{v_y}\, g\bar{n}\cos\theta$$

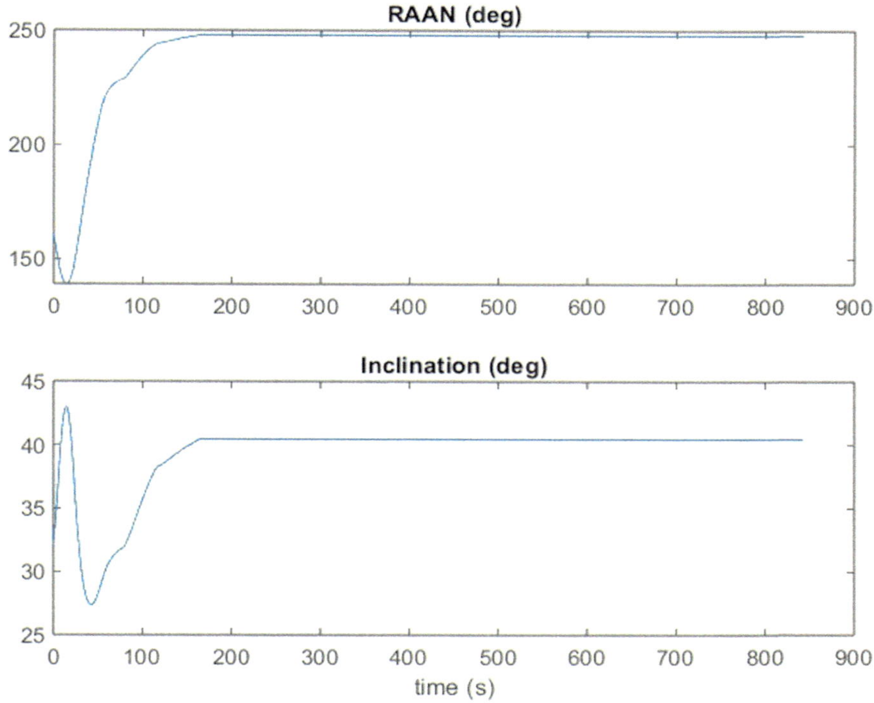

Fig. 3.16 RAAN and inclination (deg) during the flight with $t_p = 17.8$ (including coasting of 677 s)

and the cost-constraint function with constant lagrangian multipliers $\nu_1 - \nu_4$:

$$\Phi = t_f + \nu_1(x_f - r_f) + \nu_2 v_{xf} + \nu_4(v_{yf} - \sqrt{\frac{\mu}{r_f}}))$$

The Hamilton equations for the multipliers are:

$$\begin{aligned}
\dot{\lambda}_1 &= -\frac{\partial H}{\partial x} = 0 \\
\dot{\lambda}_1 &= -\frac{\partial H}{\partial y} = 0 \\
\dot{\lambda}_4 &= -\frac{\partial H}{\partial v_x} = -\lambda_1 \\
\dot{\lambda}_5 &= -\frac{\partial H}{\partial v_y} = -\lambda_2
\end{aligned} \tag{3.14}$$

where $\lambda_1 = \lambda_x$, $\lambda_2 = \lambda_y$, $\lambda_4 = \lambda_{vx}$, $\lambda_5 = \lambda_{vy}$. The solutions are:

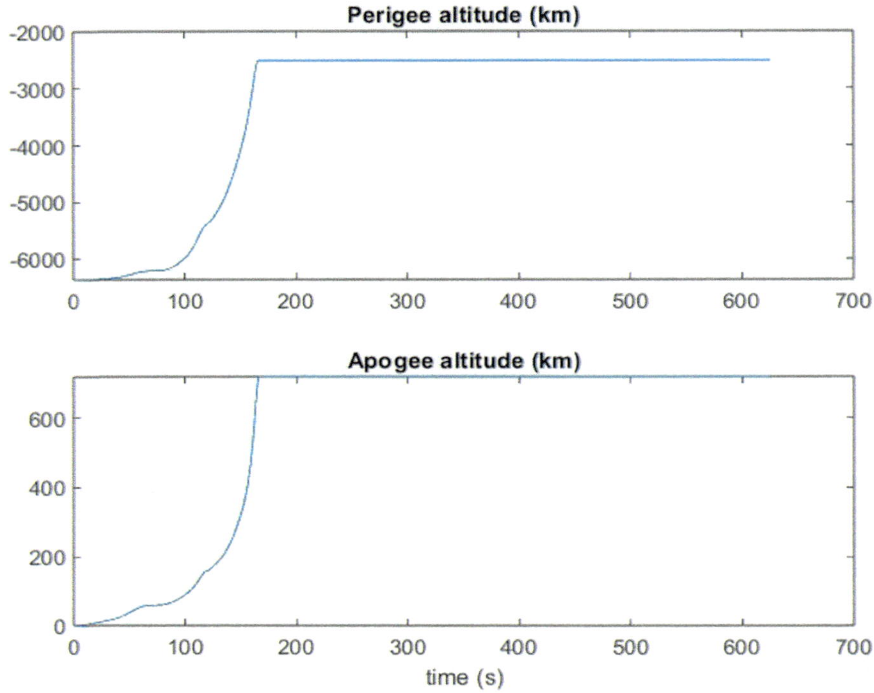

Fig. 3.17 Pericenter and apocenter altitude (km) during the flight with $t_p = 17.8$ (including coasting of 677 s)

$$\begin{cases} \lambda_1 = \lambda_{1_0} \\ \lambda_2 = \lambda_{2_0} \\ \lambda_4 = -\lambda_{1_0} t + \lambda_{4_0} \\ \lambda_5 = -\lambda_{2_0} t + \lambda_{5_0} \end{cases} \tag{3.15}$$

Since the variable y_f is left free, a condition on the final value of the multiplier λ_2 is given (see Eq. (3.9):

$$\lambda_{2f} = -\frac{\partial \Phi}{\partial y_f} = 0$$

so λ_2 is identically zero and $\lambda_5 = \lambda_{5_0}$. The necessary condition for optimal guidance angle is:

$$\frac{\partial H}{\partial \theta} = 0$$

that is:

$$\lambda_4 \cos\theta - \lambda_5 \sin\theta = 0$$

Table 3.1 Data of the trajectory with $t_p = 17.8$ s and coasting duration $t_c = 677$ s

time (s)	Event	Altitude (km)	Rel. Vel (m/s)	Rel. fpa (deg)
0	Lift-off	0.0	0.0	90.0
6	End vertical arc	0.2	124	89.3
28.3	End of pitch manoeuvre	4.7	384	60.0
42.2	Pitch over	12	815	36.6
80	First stage burn out	37	1549	23.5
119	Secont stage burn out	67	3839	15.9
166	Third stage burn out	127	6717	13.2
843	End of coasting	650	6369	0.0

and the optimal guidance verifies the tangent linear law:

$$\tan\theta = \frac{\lambda_4}{\lambda_5} = \frac{\lambda_{1_0} t + \lambda_{4_0}}{\lambda_{5_0}}$$

The solutions are:

$$\sin\theta = \frac{\lambda_4}{\sqrt{\lambda_4^2 + \lambda_5^2}}, \quad \cos\theta = \frac{\lambda_5}{\sqrt{\lambda_4^2 + \lambda_5^2}}$$

that is

$$\sin\theta = \frac{\frac{\lambda_{4_0}}{\lambda_{5_0}} - \frac{\lambda_{1_0}}{\lambda_{5_0}} t}{\sqrt{(\frac{\lambda_4}{\lambda_{5_0}})^2 + 1}}, \quad \cos\theta = \frac{sign(\lambda_{5_0})}{\sqrt{(\frac{\lambda_4}{\lambda_{5_0}})^2 + 1}} \tag{3.16}$$

where the sign = ±1 ambiguity will be solved later.

The guidance law depends on the two parameters $\lambda_{1_0}/\lambda_{5_0}$, $\lambda_{4_0}/\lambda_{5_0}$, here renamed as λ_{1_0}, λ_{4_0}. Inserting the guidance (3.16) in the equation of motion:

$$\dot{v}_x = g\bar{n}\frac{\lambda_{4_0} - \lambda_{1_0} t}{\sqrt{(\lambda_{4_0} - \lambda_{1_0} t)^2 + 1}} - g$$

$$\dot{v}_y = g\bar{n}\frac{sign(\lambda_{5_0})}{\sqrt{(\lambda_{4_0} - \lambda_{1_0} t)^2 + 1}}$$

with $a_1 = \lambda_{4_0}$, $a_2 = -\lambda_{1_0}$, it is possible to integrate $\dot{v}_x$:

$$v_x(t) = v_x(t_0) + g\bar{n} \int_{t_0}^{t} \frac{a_1 + a_2\tau}{\sqrt{(a_1 + a_2\tau)^2 + 1}} \, d\tau - g(t - t_0)$$

That is

$$v_x(t) = v_x(t_0) + g\frac{\bar{n}}{a_1} \left[\sqrt{a_1 + a_2\tau}\right]_{t_0}^{t} - g(t - t_0)$$

Then it is possible to derive x by integration of v_x:

$$x(t) = x(t_0) + vx(_t0)(t - t_0) - \frac{g}{2}(t - t_0)^2 + g\frac{\bar{n}}{a_1} \int_{t_0}^{t} \sqrt{a_1 + a_2\tau} \, dt$$

where

$$\int \sqrt{a_1 + a_2\tau} \, dt = \frac{log(a_1 + a_2t + \sqrt{1 + (a_1 + a_2\tau)^2})}{2a_2} + \frac{(a_1 + a_2t)\sqrt{a_1 + a_2t}}{2a_2}$$

Integrating $\dot{v}_y$:

$$v_y(t) = v_y(t_0) + c_1 \int_{t_0}^{t} [a_1 + a_2\tau]^{-1/2} \, d\tau$$

that is

$$v_y(t) = v_y(t_0) + \frac{c_1}{a_2}[asinh(a_1 + a_2\tau]_{t_0}^{t}$$

with $c_1 = sign(\lambda_{5_0})\, g\, \bar{n}$. Then three equations are determined by the required final conditions:

$$x(t_f) = r_f \tag{3.17}$$

$$v_x(t_f) = 0 \tag{3.18}$$

$$v_y(t_f) = \sqrt{\mu/r_f} \tag{3.19}$$

in the unknowns a_1, a_2, c_1.

The ambiguity of sign in c_1 can be solved returning to the relationships among the control angle θ and the lagrangian multipliers. From the formulas of $\sin\theta$, $\cos\theta$ one has at time $t_0 = 0$:

$$\sin\theta_0 = \frac{\lambda_{4_0}}{\sqrt{\lambda_{4_0}^2 + 1}}$$

$$\cos\theta_0 = \frac{sign(\lambda_{5_0})}{\sqrt{\lambda_{4_0}^2 + 1}}$$

then

$$\tan\theta_0 = sign(\lambda_{5_0})\, \lambda_{4_0} \tag{3.20}$$

and at final time t_f:

$$\tan\theta_f = \frac{\lambda_4(t_f)}{\lambda_{5_0}} = sign(\lambda_{5_0})\,(-\lambda_{1_0} t_f + \lambda_{4_0})$$

that is

$$\tan\theta_f = -sign(\lambda_{5_0})\lambda_{1_0} t_f + \tan\theta_0$$

then the final time is equal to:

$$t_f = -sign(\lambda_{5_0})\frac{\tan\theta_f - \tan\theta_0}{\lambda_{1_0}}$$

The sign ambiguity $sign(\lambda_{5_0})$ is now resolved. since t_f must be greater than zero.

Denote the correct $sign(\lambda_{5_0})$ by ϵ, then the final condition on the velocity $v_y(t_f)$ generates a formula for λ_{1_0}:

$$\sqrt{\frac{\mu}{r_f}} = v_y(t_0) + \epsilon\frac{g\bar{n}}{a_2}(asinh(a_1 + a_2 t_f) - asinh(a_1))$$

that is:

$$\sqrt{\frac{\mu}{r_f}} = v_y(t_0) + \epsilon\frac{g\bar{n}}{a_2}[asinh(\tan\theta_f) - asinh(\tan\theta_0)]$$

Recalling that $a_2 = -\lambda_{1_0}$:

$$\lambda_{1_0} = -\epsilon\frac{g\bar{n}}{\sqrt{\frac{\mu}{r_f}} - v_y(t_0)}[asinh(\tan\theta_f) - asinh(\tan\theta_0)] \quad (3.21)$$

It follows from the above equation for t_f:

$$t_f = \epsilon\frac{\sqrt{\frac{\mu}{r_f}} - v_y(t_0)}{g\bar{n}}\frac{\tan\theta_f - \tan\theta_0}{asinh(\tan\theta_f) - asinh(\tan\theta_0)} \quad (3.22)$$

The guessed values $\theta_0 = \gamma_0, \theta_f = 0$ are used to define the unknowns $\lambda_{4_0}, \lambda_{1_0}, t_f$ by the formulas (3.20), (3.22), (3.21). According to these values the final conditions (3.17), (3.18), (3.19) must be fulfilled: if not the guessed values are updated. Of course it may happen that no values θ_0, θ_f for any $t_f < t_b$ (engine burn time) allow the final conditions to be met. In such a case the initial values x_0, vx_0, vy_0 at the end of the coasting are not suitable.

The time optimal guidance is applied to the Scout trajectory having the apogee at $h = 650$ km reached after the burn of the first three stages and coasting time $t_c = 677$ s, see the trajectory data in Figs. 3.12, 3.13, 3.14, 3.15, 3.16 and 3.17. The guidance starts at time $t = 740$ s from lift-off, when the flight path angle has the

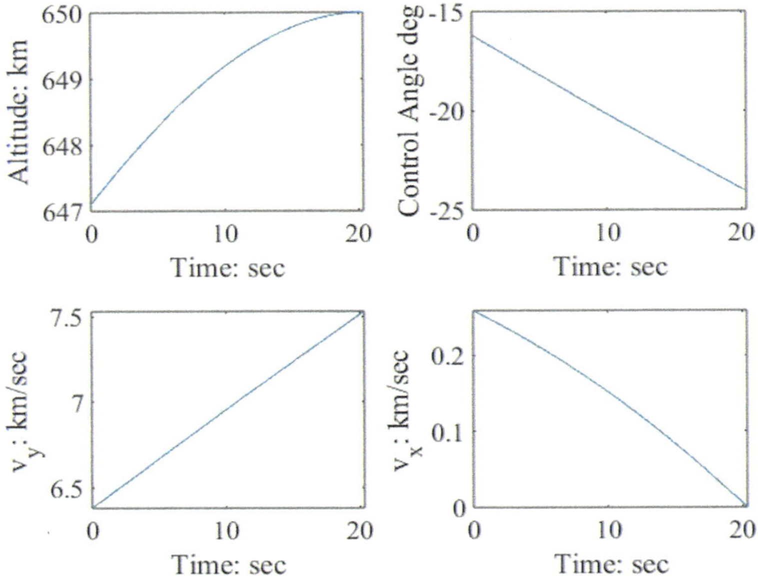

Fig. 3.18 Altitude, guidance angle θ, tangential and radial velocities achieved by optimal time guidance applied to the fourth stage of Scout launcher

value $\gamma_0 = 2.33$ deg, as in Problem 3.1. The payload mass m_{pay} including payload adapter is 150 kg. The results of thrusting the fourth stage of Scout according to the optimal time guidance are in Fig. 3.18. The the injection into orbit is achieved: the altitude is $h = 650$ km, there is no radial velocity and the tangential velocity is equal to the orbital velocity 7.5 km/s. The minimum time to achieve the target orbit is $t_{opt} = 20$ s, see Problem 3.1.

Such a time is lower than the burn out time t_{b_4} and a propellant mass equal to $m_p = m_{d4} * (t_{b_4} - t_{opt})$ can in principle be saved. For the Scout vehicle $m_{d_4} = 8.4$ Kg/s, $t_{b_4} = 34$ s, $t_{opt} = 20$ s. Then the propellant saved would be equal to $m_p = 117.5$ kg. Of course a stage with solid propellant can not be shut before the burn time, and it is possible to investigate whether this extra propellant mass can be added to the original payload mass $m_{pay} = 150$ kg. In fact Problem 3.3 shows that the payload of 250 kg is the maximum payload that can be injected into the target orbit (circular, $h = 650$ km, $i = 49$ deg) by the Scout G launcher from the San Marco Platform.

3.5 Q-Guidance

After the atmospheric phase of the trajectory, the launcher is free to flight with incidence angle and the trajectory can be optimized and an optimal nominal trajectory is determined. However it must be taken into account that several perturbations can produce deviations from the nominal trajectory so that the guidance of the launcher must be implemented in real time to compare the real state of the system with the nominal one. If the difference between real and nominal state is too large, then two different strategies can be adopted.

The first strategy is to return to the nominal trajectory within the time interval of control: this kind of guidance is called **perturbative guidance** and its task is tracking the nominal trajectory by correcting the deviations from the nominal state. A different approach is the **adaptive guidance** where the system determines a new reference trajectory to be followed starting from the current position to the final target. In the adaptive guidance the updated trajectories are not optimal but possibly "near optimal" trajectories. This is due to the need to generate new trajectories in real time at cost of a reduced accuracy in the mathematical model and optimality conditions.

Generally perturbative guidance algorithms require a large number of data to be stored in the on board computer, since the nominal trajectory data and derivative terms must be at disposal and used at several check points. On the other hand the amount of operations to be performed is limited, since the control commands are obtained by linear combinations of the nominal states and of the state errors with respect to the nominal trajectory. As opposite to perturbative guidance, adaptive guidance does not need many data to be stored, but requires a greater number of operations to be performed by the on board computer.

Adaptive guidance algorithms are mostly used in current launchers and two types of adaptive guidance will be described: the first is the Q-guidance, the second is the iterative tangent linear algorithm.

The Q-guidance is one of the first guidance algorithms used in space: it is an evolution of algorithms used for the missile interception (in flat Earth models) and was used in the Apollo missions. A very interesting presentation of the guidance, also from the historical point of view, is given in the book of Battin [8]. The Q-guidance is based on the concept of the "velocity to be gained". This is the velocity that a spacecraft at $\mathbf{r}_0$ must have to reach ballistically (that is along a non propelled arc of the trajectory) a required position $\mathbf{r}_1$. As the position $\mathbf{r}_1$ is approached, this guidance "in position" is shifted into a guidance "in velocity" to achieve the required orbital velocity.

The velocity to be gained concept is related to the propellant saving since the strategy is to perform a no-thrust arc switching off the engine as soon as possible. On the other hand it is not possible to claim that the guidance is optimal but it is easy to be implemented on board as a real time guidance [9]. In the following the Q-guidance will be presented in different versions; the guidance in position will be first introduced in the hypothesis of flat Earth and impulsive thrust. This makes it easier to understand the concept of velocity to be gained and to derive the case of continuous thrust. Then impulsive and continuous thrust will be described for the spherical Earth model. The guidance in velocity (impulsive and continuous thrust) will be limited to the flat Earth case, which is simpler and highly representative of the space launcher trajectory at the final phase of orbit injection.

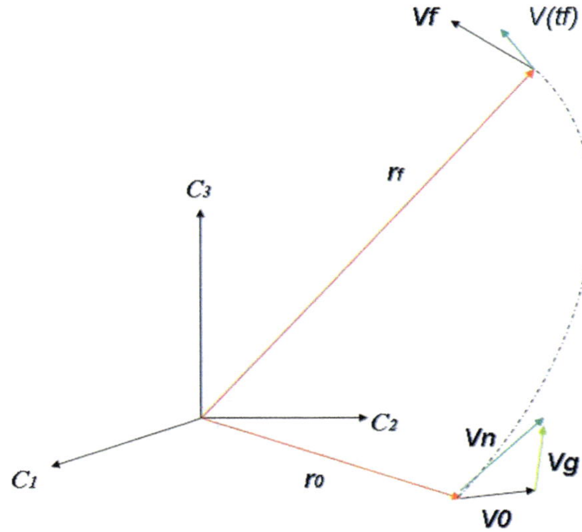

Fig. 3.19 Current velocity and velocity to be gained for a transfer from $\mathbf{r}_0$ to $\mathbf{r}_f$

Guidance in Position, Flat Earth

The problem is the following: a spacecraft in position $\mathbf{r}_0$ with velocity $\mathbf{V}_0$ must be guided to reach a target position $\mathbf{r}_f$. The idea is to transfer the spacecraft on a ballistic trajectory connecting the positions $\mathbf{r}_0$ and $\mathbf{r}_f$. There are an infinite number of ballistic trajectories connecting two points each of which corresponds to a different velocity $\mathbf{V}_n$: these velocities are called "necessary velocities" to transfer the spacecraft from its current state to the target position. According to the mission and the propulsion engine requirements, one of the necessary velocities will be selected and defined as the "required velocity" $\mathbf{V}_R$. The velocity $\mathbf{V}_g$ necessary to pass from the current velocity $\mathbf{V}_0$ to the required velocity $\mathbf{V}_R$ is the called "velocity to be gained" $\mathbf{V}_g$ (Fig. 3.19):

$$\mathbf{V}_g = \mathbf{V}_R - \mathbf{V}_0 \tag{3.23}$$

In the case of flat Earth and impulsive thrust the problem can be modeled by the equations of motion

$$\begin{cases} \dfrac{d\mathbf{r}}{dt} = \mathbf{V} \\[2ex] \dfrac{d\mathbf{V}}{dt} = -g\,\hat{h} \end{cases}$$

and the problem is to modify the initial velocity $\mathbf{V}_0$ to reach the position $\mathbf{r}_f$.

If the current spacecraft conditions are $\mathbf{r}_0$, $\mathbf{V}_0$ integration of the equations of motion gives:

$$\begin{cases} \mathbf{V} = -g\,(t - t_0)\,\hat{h} + \mathbf{V}_0 \\ \mathbf{r} = -g\,\dfrac{(t - t_0)^2}{2}\,\hat{h} + \mathbf{V}_0\,(t - t_0) + \mathbf{r}_0 \end{cases} \tag{3.24}$$

If at final time t_f the position of the spacecraft must be equal to $\mathbf{r}_f$, one must have:

$$\mathbf{r_f} = -g\,\frac{(t_f - t_0)^2}{2}\,\hat{h} + \mathbf{V}_n\,(t_f - t_0) + \mathbf{r}_0$$

where the velocity in the formula is not longer the current $\mathbf{V}_0$ but it must be one of the necessary velocities $\mathbf{V}_n$. Therefore the necessary velocities have the equation:

$$\mathbf{V}_n = \frac{\mathbf{r_f} - \mathbf{r_0}}{t_f - t_0} + \frac{g}{2}(t_f - t_0)\,\hat{h} \tag{3.25}$$

As the final time t_f varies, there are infinitely many necessary velocities with components $(\dot{X}_n, \dot{Y}_n, \dot{Z}_n)$. Choosing (X, Z) as type plane of the trajectory, the two components of the vector in (3.25) are:

$$\begin{cases} \dot{X}_n = \dfrac{X_f - X_0}{t_f - t_0} \\ \dot{Z}_n = \dfrac{Z_f - Z_0}{t_f - t_0} + g\,\dfrac{(t_f - t_0)}{2} \end{cases}$$

Deriving the time $t_f - t_0 = (X_f - X_0)/\dot{X}_n$, one has the *curve of the necessary velocities*:

$$\dot{Z}_n = \frac{Z_f - Z_0}{X_f - X_0}\,\dot{X}_n + \frac{g}{2}(X_f - X_0)\frac{1}{\dot{X}_n} \tag{3.26}$$

Note that for high values of $\dot{X}_n$, that is low values of the final time t_f, the curve is asymptotic to a straight line, whereas for low values of $\dot{X}_n$, that is high values of the final time t_f, the curve is asymptotic to an hyperbola, as depicted in Fig. 3.20. Each point of the curve corresponds to a velocity that allows the transfer of the spacecraft from $\mathbf{r}_0$ to $\mathbf{r}_f$, and different velocities correspond to different transfer times.

If the current velocity $\mathbf{V}_0$ is shown on the same plot of the curve of the necessary velocities, then any velocity $\mathbf{V}_g$ connecting $\mathbf{V}_0$ to the curve can be a velocity to be gained. In Fig. 3.20 it is shown the case of minimum $\mathbf{V}_g$, which is the minimum distance of $\mathbf{V}_0$ from the curve of the necessary velocities. Note that the direction of the impulsive thrust is just the direction of the velocity to be gained reported in the Figure.

More generally, if ΔV is the variation of velocity that the propulsion system can offer, the velocities that the launcher can achieve are those given by the sum of $\mathbf{V}_0$ and the variation of velocity ΔV as in Fig. 3.21 where ΔV is the radius of a circle

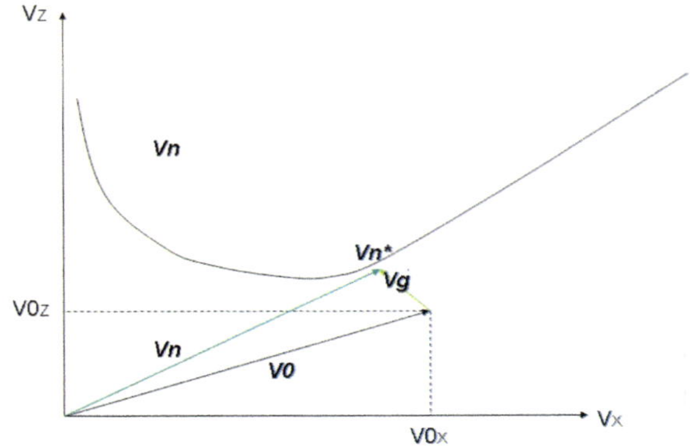

Fig. 3.20 The curve of the necessary velocities and the velocity to be gained $\mathbf{V}_g$ in the case of minimum ΔV

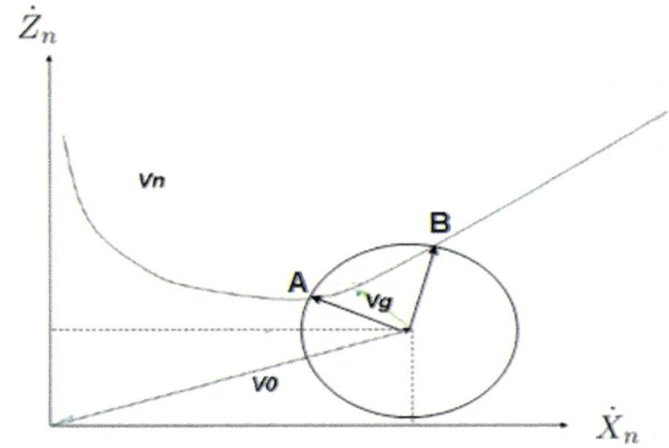

Fig. 3.21 The curve of the necessary velocities and the available $\mathbf{V}_g$

centered in $\mathbf{V}_0$. In this way an arc of necessary velocities can be targeted: if the thrust is by solid propellant then all the ΔV must be used and there are only two possible "necessary velocities" indicated by A and B in Fig. 3.21. If the point A is chosen as the required velocity, the trajectory will be more direct and faster while choosing B will produce a loft trajectory to $\mathbf{r}_f$, see Fig. 3.22.

In summary, once the required velocity $\mathbf{V}_R$ is selected, the modulus and direction of the velocity to be gained is given: this corresponds to the impulsive thrust that must be given to the spacecraft to reach $\mathbf{r}_f$ along a ballistic arc.

In the case of continuous thrust the equations of motion are:

$$\frac{d\mathbf{V}}{dt} = -g\,\hat{h} + \frac{\mathbf{T}}{m} \tag{3.27}$$

and the velocity to be gained will change on time since both the current and the required velocities are changing during the thrust phase. Let us compute this variation by differentiating the definition of $\mathbf{V}_g$

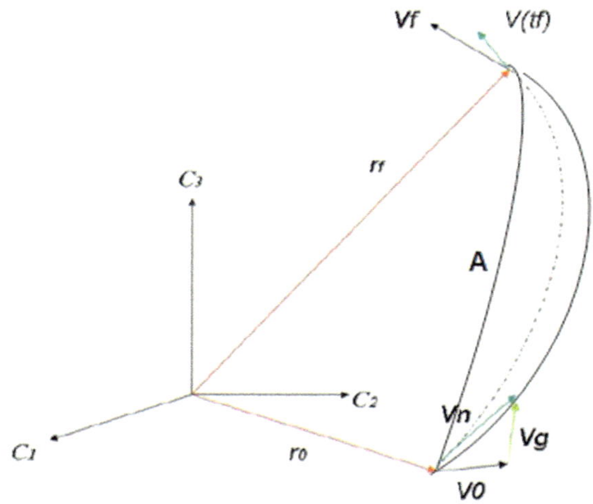

Fig. 3.22 The trajectories corresponding to the required velocities A and B

$$\frac{d}{dt}\mathbf{V}_g = \frac{d}{dt}(\mathbf{V}_R - \mathbf{V}) \tag{3.28}$$

In (3.28) the subscript index 0 in the definition of the velocity is not given, since the current velocity is changing in time. Moreover, having fixed a final time t_f (hence a required velocity among the necessary velocities), $\mathbf{V}_R$ in (3.28) is the required velocity at the current position $\mathbf{r}(t)$ to reach $\mathbf{r}_f$ at time t_f. Then $\mathbf{V}_R$ is equal to

$$\mathbf{V}_R = \frac{\mathbf{r_f} - \mathbf{r}(t)}{t_f - t} + \frac{g}{2}(t_f - t)\,\hat{h} \tag{3.29}$$

Deriving Eq. (3.29) one has:

$$\frac{d\mathbf{V}_R}{dt} = \frac{\mathbf{r_f} - \mathbf{r}(t)}{(t_f - t)^2} - \frac{\frac{d\mathbf{r}}{dt}}{(t_f - t)} - \frac{g}{2}\,\hat{h}$$

where $\frac{d\mathbf{r}}{dt} \doteq \mathbf{V}$, then:

$$\frac{d\mathbf{V}_g}{dt} = \frac{d\mathbf{V}_R}{dt} - \frac{d\mathbf{V}}{dt} = \left(\frac{\mathbf{r_f} - \mathbf{r}(t)}{(t_f - t)^2} - \frac{\mathbf{V}}{t_f - t} - \frac{g}{2}\,\hat{h}\right) - \left(-g\,\hat{h} + \frac{\mathbf{T}}{m}\right)$$

that is:

$$\frac{d\mathbf{V}_g}{dt} = \left(\frac{\mathbf{r_f} - \mathbf{r}(t)}{(t_f - t)^2} - \frac{\mathbf{V}}{t_f - t} + \frac{g}{2}\,\hat{h}\right) - \frac{\mathbf{T}}{m}$$

which is equal to

$$\frac{d\mathbf{V}_g}{dt} = \frac{1}{t_f - t}\left(\frac{\mathbf{r_f} - \mathbf{r}(t)}{(t_f - t)} + \frac{g}{2}(t_f - t)\,\hat{h}\right) - \frac{\mathbf{V}}{t_f - t} - \frac{\mathbf{T}}{m}$$

By definition of $\mathbf{V}_R$, see Eq. (3.29), one has:

$$\frac{d\mathbf{V}_g}{dt} = \frac{\mathbf{V}_R - \mathbf{V}}{t_f - t} - \frac{\mathbf{T}}{m}$$

that is

$$\frac{d\mathbf{V}_g}{dt} = \frac{\mathbf{V}_g}{t_f - t} - \frac{\mathbf{T}}{m}. \tag{3.30}$$

This is the differential equation for the velocity to be gained, depending only on the final time t_f and the thrust $\mathbf{T}$. Suppose, as in the impulsive case, that the thrust is directed as the velocity to be gained, that is $\mathbf{T} = T\,\hat{V}_g$. Then Eq. 3.30 is:

$$\frac{d\mathbf{V}_g}{dt} = \left(\frac{V_g}{t_f - t} - \frac{T}{m}\right)\hat{V}_g. \tag{3.31}$$

A first result that can be derived by Eq. 3.31 is that the direction of the velocity to be gained <u>does not</u> change with time, namely:

$$\frac{d\mathbf{V}_g}{dt} \doteq \dot{V}_g\hat{V}_g + V_g\frac{d\hat{V}_g}{dt}$$

and comparing with Eq. 3.31 it follows:

$$\frac{d\hat{V}_g}{dt} = 0 \qquad \Rightarrow \qquad \hat{V}_g = cost = \hat{V}_{g0}$$

and

$$\dot{V}_g = \frac{V_g}{t_f - t} - \frac{T}{m}. \tag{3.32}$$

That is, the direction of the velocity to be gained V_g has the same direction assumed at initial time t_0,which is the $\hat{V}_{g0}$ computed in the impulsive case. The direction of the thrust is then determined. As far as the thrust intensity is concerned, some considerations follow from the Eq. 3.32: it is necessary that the velocity to be gained is a decreasing function of the time, then

$$\frac{V_g}{t_f - t} - \frac{T}{m} < 0.$$

Since $\frac{T}{m}$ increases with time, it is sufficient that such an inequality holds true at the initial time, that is

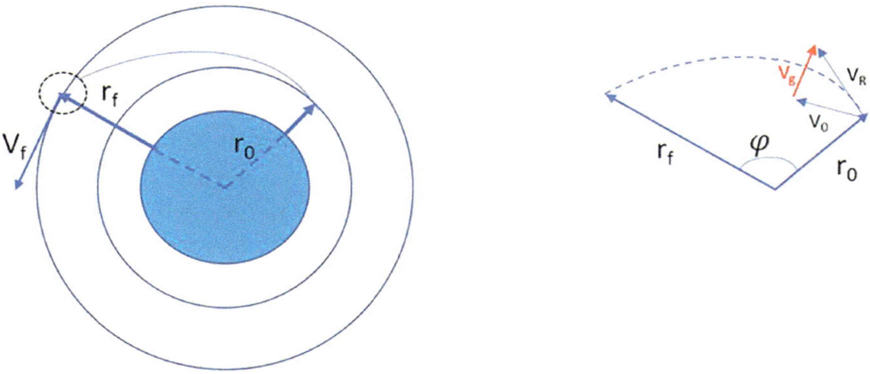

Fig. 3.23 The guidance in position in the keplerian case and the transfer angle ϕ

$$T > \frac{V_{g0}}{t_f - t_0} m(t_0) \tag{3.33}$$

and this is a requirement for the thrust intensity to reach $\mathbf{r}_f$ in the required time t_f.

Guidance in Position, Spherical Earth

The problem is the same as above, but now the ballistic trajectories to reach $\mathbf{r}_f$ are keplerian arcs, see Fig. 3.23.

At $\mathbf{r}_0$ the spacecraft must have a velocity $\mathbf{v}_0 = \mathbf{V}_R(\mathbf{r}_0)$, so to reach $\mathbf{r}_f$ ballistically, that is $(\mathbf{r}_0, \mathbf{v}_0)$ and $(\mathbf{r}_f, \mathbf{v}_f)$ are on the same keplerian orbit. Then the angular momentum and eccentricity vectors are the same in the both states:

$$\begin{aligned}
\mathbf{h}_0 &= \mathbf{r}_0 \wedge \mathbf{v}_0 = \mathbf{r}_f \wedge \mathbf{v}_f = \mathbf{h}_f \\
\mathbf{e}_0 &= -\hat{r}_0 - \frac{1}{\mu}\mathbf{h}_0 \wedge \mathbf{v}_0 = -\hat{r}_f - \frac{1}{\mu}\mathbf{h}_f \wedge \mathbf{v}_f = \mathbf{e}_f
\end{aligned}$$

From the latter equation it follows that:

$$\mu \left(\hat{r}_f - \hat{r}_0\right) + \mathbf{h}_f \wedge \mathbf{v}_f - \mathbf{h}_0 \wedge \mathbf{v}_0 = 0$$

Scalar product by $\mathbf{r}_f$ gives:

$$\mu \, \mathbf{r}_f \cdot \left(\hat{r}_f - \hat{r}_0\right) + \mathbf{r}_f \cdot \mathbf{h}_f \wedge \mathbf{v}_f - \mathbf{r}_f \cdot \mathbf{h}_0 \wedge \mathbf{v}_0 = 0$$

Consider that the mixed product is commutative:

$$\mathbf{a} \cdot (\mathbf{b} \wedge \mathbf{c}) = \mathbf{c} \cdot (\mathbf{a} \wedge \mathbf{b}) = \mathbf{b} \cdot (\mathbf{c} \wedge \mathbf{a})$$

then

$$\mu \, \mathbf{r}_f \cdot \left(\hat{r}_f - \hat{r}_0\right) + \mathbf{h}_f \cdot \mathbf{v}_f \wedge \mathbf{r}_f - \mathbf{h}_0 \cdot \mathbf{v}_0 \wedge \mathbf{r}_f = 0$$

Since $\mathbf{r}_0 \wedge \mathbf{v}_0 = \mathbf{r}_f \wedge \mathbf{v}_f$, one has the following equation for the required velocity $\mathbf{v}_0$:

$$\mu \, \mathbf{r}_f \cdot \left(\hat{r}_f - \hat{r}_0\right) + \mathbf{h}_0 \cdot \mathbf{v}_0 \wedge \mathbf{r}_0 - \mathbf{h}_0 \cdot \mathbf{v}_0 \wedge \mathbf{r}_f = 0$$

that is

$$\mu \, \mathbf{r}_f \cdot \left(\hat{r}_f - \hat{r}_0\right) + \mathbf{h}_0 \cdot \left[\mathbf{v}_0 \wedge (\mathbf{r}_0 - \mathbf{r}_f)\right] = 0$$

and finally

$$\mu \, \mathbf{r}_f \cdot \left(\hat{r}_f - \hat{r}_0\right) - (\mathbf{v}_0 \wedge \mathbf{r}_0) \cdot \left[\mathbf{v}_0 \wedge \left(\hat{r}_0 - \hat{r}_f\right)\right] = 0 \tag{3.34}$$

Let φ the transfer angle of Fig. 3.23, then:

$$\mu \mathbf{r}_f \cdot \left(\hat{r}_f - \hat{r}_0\right) = \mu r_f (1 - \cos\varphi)$$

$$\mathbf{v}_0 \wedge \mathbf{r}_0 = \left(v_r \hat{r}_0 + v_\theta \hat{\theta}_0\right) \wedge \mathbf{r}_0 = -v_\theta \, r_0 \, \hat{h}_0$$

$$\mathbf{v}_0 \wedge \mathbf{r}_f = \left(v_r \hat{r}_0 + v_\theta \hat{\theta}_0\right) \wedge r_f \left(\cos\varphi \, \hat{r}_0 + \sin\varphi \, \hat{\theta}_0\right)$$

so that

$$\mathbf{v}_0 \wedge \mathbf{r}_f = r_f v_r \sin\varphi \, \hat{h}_0 - r_f v_\theta \cos\varphi \, \hat{h}_0$$

Using the above formulas in Eq. 3.34 one has

$$\mu r_f \, (1 - \cos\varphi) + v_\theta r_0 \left(-v_\theta r_0 - r_f v_r \sin\varphi + r_f v_\theta \cos\varphi\right) = 0$$

Then, the expression of the necessary velocity in the v_r e v_θ components is derived [8]:

$$v_r = \mu \frac{1 - \cos\varphi}{r_0 \, \sin\varphi} \frac{1}{v_\theta} + \frac{r_f \, \cos\varphi - r_0}{r_f \sin\varphi} \, v_\theta \tag{3.35}$$

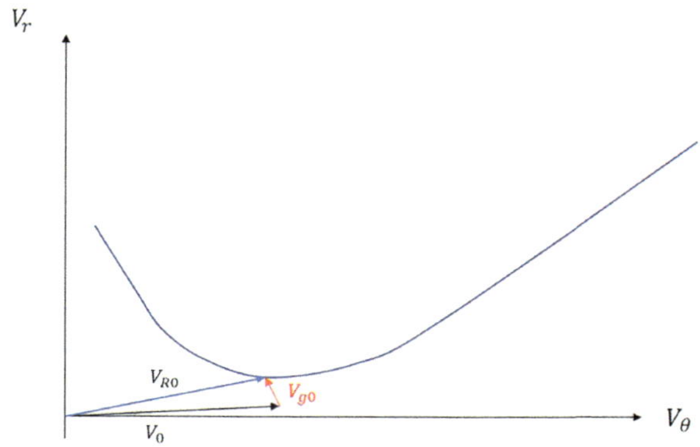

Fig. 3.24 Curve of the necessary velocities

Note that for low values of the tangential velocity v_θ the hyperbolic term in (3.35) is dominant, while for high values of v_θ the linear term in (3.35) is more important, see Fig. 3.24. Then the keplerian impulsive thrust is similar to the flat Earth case.

In the case of continuous thrust, the equation of motion is:

$$\frac{d\mathbf{V}}{dt} = -\frac{\mu}{r^3}\mathbf{r} + \frac{\mathbf{T}}{m}$$

The aim is (as before) to acquire, from the current velocity V, a velocity allowing to reach the target position $\mathbf{r}_f$ from the initial position $\mathbf{r}_0$. The velocity to be gained (after choosing a required velocity $\mathbf{V}_R$)

$$\mathbf{V}_{g_0} = \mathbf{V}_R - \mathbf{V}$$

has the differential equation

$$\begin{aligned}\frac{d\mathbf{V}_g}{dt} &= \frac{d\mathbf{V}_R}{dt} - \frac{d\mathbf{V}}{dt} = \frac{d\mathbf{V}_R}{dt} - \left(\mathbf{g}(\mathbf{r}) + \frac{\mathbf{T}}{m}\right) \\ &= \left(\frac{\partial \mathbf{V}_R}{\partial t} + \frac{\partial \mathbf{V}_R}{\partial \mathbf{r}}\frac{d\mathbf{r}}{dt}\right) - \left(\mathbf{g}(\mathbf{r}) + \frac{\mathbf{T}}{m}\right) \\ &= \left(\frac{\partial \mathbf{V}_R}{\partial t} + \frac{\partial \mathbf{V}_R}{\partial \mathbf{r}}\mathbf{V}\right) - \left(\mathbf{g}(\mathbf{r}) + \frac{\mathbf{T}}{m}\right) \\ &= \frac{\partial \mathbf{V}_R}{\partial t} + \frac{\partial \mathbf{V}_R}{\partial \mathbf{r}}(\mathbf{V}_R - \mathbf{V}_g) - \left(\mathbf{g}(\mathbf{r}) + \frac{\mathbf{T}}{m}\right) \\ &= \left(\frac{\partial \mathbf{V}_R}{\partial t} + \frac{\partial \mathbf{V}_R}{\partial \mathbf{r}}\mathbf{V}_R\right) - \frac{\partial \mathbf{V}_R}{\partial \mathbf{r}}\mathbf{V}_g - \left(\mathbf{g}(\mathbf{r}) + \frac{\mathbf{T}}{m}\right)\end{aligned}$$

In the last equation there is the acceleration along the trajectory ($\mathbf{r}_R$) related to the required velocity $\mathbf{V}_R$: this is a ballistic arch having no thrust:

$$\left(\frac{\partial \mathbf{V}_R}{\partial t} + \frac{\partial \mathbf{V}_R}{\partial \mathbf{r}}\mathbf{V}_R\right) \doteq \frac{d\mathbf{V}_R(\mathbf{r}_R)}{dt} = \mathbf{g}(\mathbf{r}_R)$$

Then

$$\frac{d\mathbf{V}_g}{dt} = \mathbf{g}(\mathbf{r}_R) - \mathbf{g}(\mathbf{r}) - \frac{\partial \mathbf{V}_R}{\partial \mathbf{r}}\mathbf{V}_g - \frac{\mathbf{T}}{m}$$

As for the two gravitational terms, at current time the gravity is the same in the two trajectories: namely the trusted and the ballistic and the two vectors of the position correspond to the same point ($\mathbf{r} = \mathbf{r}_R$), so

$$\mathbf{g}(\mathbf{r}_R) = \mathbf{g}(\mathbf{r})$$

hence the velocity to be gained has the final equation:

$$\frac{d\mathbf{V}_g}{dt} = -\frac{\partial \mathbf{V}_R}{\partial \mathbf{r}}\mathbf{V}_g - \frac{\mathbf{T}}{m} \doteq -[Q]\mathbf{V}_g - \frac{\mathbf{T}}{m}$$

The initial condition $\mathbf{V}_{g0}$ is obtained by the impulsive case and the thrust direction can be chosen as $\hat{T} \equiv \hat{V}_g$.

Because of the matrix $[Q]$, the guidance $\hat{T} \equiv \hat{V}_g$ does not imply that the direction of the velocity to be gained is constant, as in the flat Earth case.

Guidance in Velocity, Flat Earth [10]

As usual in guidance algorithms, the Q-guidance in position has a singularity when the spacecraft reaches exactly the required position at the final time t_f, see Eq. 3.25. Then the Q-guidance must be followed until the spacecraft is sufficiently close to the final target position $\mathbf{r}_f$, and it must be switched to a guidance in velocity before reaching $\mathbf{r}_f$ to gain a required final velocity $\mathbf{V}_f$. In the case of keplerian circular orbit the required velocity $\mathbf{V}_f$ has a module equal to $\sqrt{\frac{\mu}{R_\oplus + h}}$ where h is the target altitude.

The direction of the velocity is always orthogonal to the satellite radius vector and there is no need to specify a particular injection point on the circular orbit. Since the altitude is rather close to the required one and the fourth stage burn out is generally short, it is reasonable to assume that there is a small arc (transfer angle) between the current and the injection point. This angle can be found using the following approximations. The true anomaly will change according to the equation [6]:

$$\dot{\theta} = \frac{V}{r}\cos\gamma$$

In this last phase of the trajectory, the thrust direction is chosen to increase the velocity, while the target altitude has already been reached: then the velocity V changes rapidly and in the meantime r and γ are almost constant, so:

$$\dot{\theta} \sim \frac{\cos \gamma_0}{r_0} V$$

and the variation of θ which is equal to the transfer angle ϕ is given by:

$$\Delta\theta = \phi = \frac{\cos \gamma_0}{r_0} \int_0^{t_f} V(\tau)\, d\tau \tag{3.36}$$

In a short time period we can also neglect the losses and consider only the thrust force, then

$$V = V_0 + \int_0^{t_f} \frac{T}{m} d\tau \tag{3.37}$$

The integral in (3.37) is defined "first thrust integral":

$$f_1(t) = \int_0^{t} \frac{T}{m} d\tau \tag{3.38}$$

and gives the Tziolkowski variation of velocity:

$$f_1(t) = gIsp \log \frac{m_0}{m_0 - m_d t} \tag{3.39}$$

The integral in the formula of the velocity (3.36) is then:

$$\int_0^{t_f} V(\tau)\, d\tau = \int_0^{t_f} (V_0 + f_1(t))\, dt = V_0\, t_f + \int_0^{t_f} f_1(\tau)\, d\tau$$

Defining

$$f_3(t) = \int_0^{t} f_1(\tau)\, d\tau \tag{3.40}$$

we get

$$\int_0^{t_f} V(\tau)\, d\tau = V_0\, t_f + f_3(t_f)$$

where

$$f_3(t) = g\, Isp \left[log(1 - q\, t)\, t - t - log(1 - q\, t) \frac{1}{q} \right] ,\ q = \frac{\dot{m}}{m_0} \tag{3.41}$$

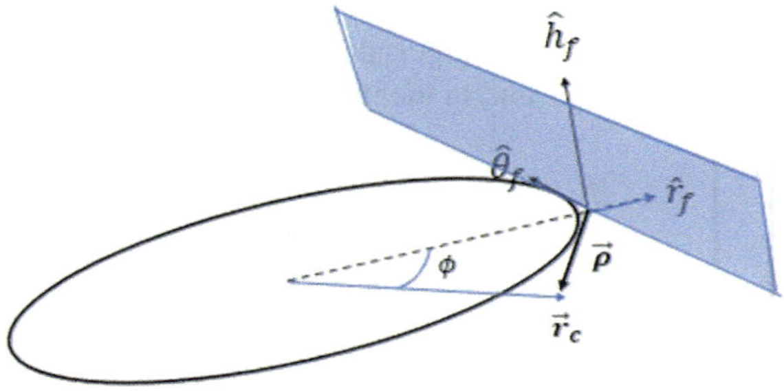

Fig. 3.25 The position $\mathbf{r}_c$ at the end of coasting, the target position $\mathbf{r}_f$ on the final orbit and the transfer angle ϕ. The flat Earth horizontal plane is shown

In Eq. (3.41) t is the thrusting time to achieve the required velocity, also called time to go T_{go}. According to the Tziolkowski law, this is equal to:

$$T_{g0} = \frac{m_0}{m_f}\,(1 - exp^{-\frac{DV}{g*Isp}}) \;\;, \quad \Delta V = V_f - V_0 \tag{3.42}$$

Then, inserting $t_f = T_{go}$ in (3.36) we have the transfer angle

$$\phi = \frac{\cos\gamma_0}{r_0}\left[V_0 T_{go} + g\,Isp\,\left(log(1 - q\,T_{go})\,T_{go} - T_{go} - log(1 - q\,T_{go})\frac{1}{q}\right)\right] \tag{3.43}$$

Different transfer angles will be obtained for different states (r_0, V_0, θ_0) along the coasting orbit.

Knowing the transfer angle we have the target injection point $\mathbf{r}_f$ and we consider the inertial reference frame TPI with fixed axes $\hat{r}_f$, $\hat{\theta}_f$, $\hat{h}_f$. The position of the last stage is defined by the vector distance $\boldsymbol{\rho} = x\hat{r}_f + y\hat{\theta}_f + z\hat{h}_f$ from the target point, see Fig. 3.25.

Flat Earth approximation is considered, so the equations of motion are

$$\frac{d\mathbf{r}}{dt} = \mathbf{V} \;\;, \quad \frac{d\mathbf{V}}{dt} = -g\hat{r}_f + \frac{\mathbf{T}}{m} \tag{3.44}$$

The rocket coordinates at the end of the coasting (the coasting trajectory is planar)

$$\mathbf{r}_c = x_c\hat{r} + y_c\hat{\theta} \;\;, \quad \mathbf{V}_c = u_c\hat{r} + v_c\hat{\theta}$$

are transformed to TPI coordinates (x_0, y_0), (u_0, v_0) by the rotation of the transfer angle ϕ:

$$x_0 = x_c\cos(\phi) + y_c\sin(\phi) \;, \quad y_0 = -x_c\sin(\phi) + y_c\cos(\phi)$$

$$u_0 = u_c\cos(\phi) + v_c\sin(\phi) \;, \quad v_0 = -u_c\sin(\phi) + v_c\cos(\phi)$$

As it was done in the case of guidance in position, the impulsive case is studied first:

(i) a velocity $\mathbf{V}_{Req} = (u_R, v_R)$ at the end of coasting is required to achieve the final velocity $V_f \hat{\theta}_f$ after the completion of the transfer angle ϕ

(ii) an impulsive variation of velocity $\mathbf{V}_g$ (velocity to be gained) is provided to the rocket so that

$$\mathbf{V}_c + \mathbf{V}_g = \mathbf{V}_{Req}$$

where $\mathbf{V}_c$ is the velocity at the end of the coasting and the required end conditions are achieved by a ballistic arc.

In the impulsive case one has the equations

$$\begin{cases} \dot{x} = u \ , \quad \dot{y} = v \\ \dot{u} = -g \ , \quad \dot{v} = 0 \end{cases} \tag{3.45}$$

and the solution:

$$\mathbf{V}(t) = -g(t - t_c)\hat{r}_f + \mathbf{V}_c$$

$$\mathbf{r}(t) = -\frac{g}{2}(t - t_c)^2\, \hat{r}_f + \mathbf{V}_c(t - t_c) + \mathbf{r}_c$$

Supposing that the time $t = \tilde{t}$ is the time when the final velocity is achieved starting from the required velocity $\mathbf{V}_R$ instead of the current velocity $\mathbf{V}_c$, then

$$\mathbf{V}(\tilde{t}) = -g(\tilde{t} - t_c)\hat{r}_f + \mathbf{V}_{Req} = \mathbf{V}_f = V_f \hat{\theta}_f$$

so the radial and tangential components of $\mathbf{V}_R$ are equal to:

$$u_R = g(\tilde{t} - t_c) \tag{3.46}$$

$$v_R = V_f \tag{3.47}$$

From the equation of motion, the function

$$\frac{u^2}{2} + g\, x$$

is constant so its value does not change from the initial to the final condition:

$$\frac{u_R^2}{2} + g\, x_0 = \frac{u_f^2}{2} + g\, x_f = g\, r_f$$

where the last equality follows from the fulfillment of the final conditions (zero radial velocity $u_f = 0$ and distance $x_f = r_f$).

Then

$$u_R = \sqrt{2\, g(r_f - x_0)} \tag{3.48}$$

Equations (3.48), (3.47) determine the required velocity, and the velocity to be gained by impulsive thrust is:

$$\mathbf{V}_{g0} = (\sqrt{2g(r_f - x_0)} - u_0,\, V_f - v_0) \tag{3.49}$$

Now the continuous thrust is considered, and the derivative on time of the velocity to be gained is evaluated:

$$\frac{d\mathbf{V}_g}{dt} = \frac{d\mathbf{V}_{Req}}{dt} - \frac{d\mathbf{V}}{dt} = \frac{d\mathbf{V}_{Req}}{dt} - (\frac{\mathbf{T}}{m} - g\hat{r}_f) \tag{3.50}$$

On the other hand

$$\frac{d\mathbf{V}_{Req}}{dt} = \dot{u}_R\hat{r}_f + \dot{v}_R\hat{\theta}_f = \sqrt{\frac{g}{2}}\frac{-u}{\sqrt{r_f - x}}\hat{r}_f$$

Since $u = u_R - u_g$, one has

$$\frac{d\mathbf{V}_{Req}}{dt} = \sqrt{\frac{g}{2}}\frac{u_g - u_R}{\sqrt{r_f - x}}\hat{r}_f$$

and, using the formula (3.48):

$$\frac{d\mathbf{V}_{Req}}{dt} = (\sqrt{\frac{g}{2}}\frac{u_g}{\sqrt{r_f - x}} - g)\hat{r}_f$$

As a result, (3.50) corresponds to:

$$\frac{d\mathbf{V}_g}{dt} = \sqrt{\frac{g}{2}}\frac{u_g}{\sqrt{r_f - x}} - \frac{\mathbf{T}}{m} \tag{3.51}$$

The direction of the thrust is taken equal to the direction of the velocity to be gained: $\hat{T} = -\hat{V}_g$. If (l_x, l_y, l_z) are the direction cosines of the thrust direction $\hat{T}$, then

$$l_x = \frac{u_g}{V_g}\,,\; l_y = \frac{v_g}{V_g}\,,\; l_z = 0\,,\; V_g = \sqrt{u_g^2 + v_g^2}$$

It follows that the Eq. (3.51) of the velocity to be gained produces the two scalar equations:

$$\begin{cases} \dot{u}_g = [\sqrt{\frac{g}{2}}\frac{1}{\sqrt{r_f - x}} - \frac{T}{V_g m}]\,u_g \\ \dot{v}_g = -\frac{T}{V_g m}\,v_g \end{cases} \tag{3.52}$$

The boosted trajectory together with the thrust direction are obtained by the Eqs. (3.44), (3.52) with the initial conditions (x_0, y_0, u_0, v_0) and (3.49).

3.6 Real Time Tangent Linear Law

Current European launchers use a real time tangent linear law to guide the last stage, from the end of the coasting phase to orbit injection.

The guidance is adaptive and the trajectory splits into several arcs with a frequency of a few seconds. The state data of the launcher and the guidance parameters are updated step by step. At each step a constant value of gravity is assumed to simplify the analysis and make it suitable for real time algorithms. In particular the in plane and the out of plane motion of the last stage are decoupled and at each step the guidance depends only on two parameters that are analytically derived imposing the required end conditions (insertion into the target orbit).

The thrust direction is determined by an out of plane angle η and an in plane angle σ both having a tangent linear law with respect to time: this law was proved to be optimal in Sect. 3.4.

The launcher state, obtained for instance by the navigation system in ECI frame is projected into the Target Plane defined by the RAAN angle Ω_T and inclination i_T. The projection is realized by the usual 3-1 sequence: $R_{ROT} = R^1_{i_T} R^3_{\omega_T}$ and state variables:

$$\begin{bmatrix} x_0 \\ y_0 \\ z_0 \end{bmatrix} = R_{ROT} \begin{bmatrix} X_0 \\ Y_0 \\ Z_0 \end{bmatrix}$$

$$\begin{bmatrix} u_0 \\ v_0 \\ w_0 \end{bmatrix} = R_{ROT} \begin{bmatrix} VX_0 \\ VY_0 \\ VZ_0 \end{bmatrix}$$

The projection into the target plane is $(x_0, y_0, z_0, u_0, v_0, w_0) \longrightarrow (x_0, y_0, u_0, v_0)$ where the out of plane variables (z_0, w_0) are expected to be small due to the previous guidance strategy, in particular the choice of azimuth angle and launch time.

The angle ϕ_0 between the projection of the launcher position on the target plane and the nodal line of the target plane is obtained by

$$sin(\phi_0) = \frac{x_0}{r_0}, \quad cos(\phi_0) = \frac{y_0}{r_0}, \quad r_0 = \sqrt{x_0^2 + y_0^2}$$

The aim is to thrust in such a way that, after a time T_{go} to be defined, the launcher reaches the target radius distance r_T with the corresponding circular velocity $V_T = \sqrt{\frac{\mu}{r_T}}$. The position on the target plane is identified by the radius r_T and the angle θ from the current $\mathbf{r}_0$ to the injection into orbit. Such an angle is approximated during the boost phase by integration from time 0 to T_{go} of the equation (3.43):

$$\Phi = \phi_0 + \frac{\cos \gamma_0}{r_0}(V_0 T_{go} + f_1(T_{go})) \tag{3.53}$$

The three rotations 3-1-3 of angles Ω_T, i_T, Φ define the direction $\hat{r}_T$ of the injection point and determine the orbital frame $(\hat{r}_T, \hat{\theta}_T, \hat{h}_T)$. This is considered as inertial during the guidance step and the equation of motion are:

$$\begin{cases} \dfrac{dx}{dt} = u \\ \dfrac{dy}{dt} = v \\ \dfrac{dz}{dt} = w \\ \dfrac{du}{dt} = \dfrac{T}{m} \cos\eta \sin\sigma + G_X(x, y, z) \\ \dfrac{dv}{dt} = \dfrac{T}{m} \cos\eta \cos\sigma + G_Y(x, y, z) \\ \dfrac{dw}{dt} = \dfrac{T}{m} \sin\eta + G_Z(x, y, z) \end{cases} \tag{3.54}$$

where (G_x, G_y, G_z) is the projection of the gravity acceleration and η, σ are the out of plane and in plane components of the thrust direction.

Average Gravity

The average gravity between the position $\mathbf{r}_0$ and the final position $\mathbf{r}_f$ is simply:

$$\mathbf{g} = \frac{1}{2}\left(-\frac{\mu}{r_f^3}\mathbf{r}_f - \frac{\mu}{r_0^3}\mathbf{r}_0\right) = \overline{G}_x\hat{r}_T + \overline{G}_y\hat{\theta}_T + \overline{G}_z\hat{h}_T \tag{3.55}$$

a constant vector in the guidance step $[\mathbf{r}_0, \mathbf{r}_f]$

If the average gravity is considered, the out of plane has an independent dynamics: OUT OF PLANE

$$\begin{cases} \dfrac{dz}{dt} = w \\ \dfrac{dw}{dt} = \dfrac{T}{m} \sin\eta + \overline{G}_z \end{cases} \tag{3.56}$$

IN PLANE

$$\begin{cases} \dfrac{dx}{dt} = u \\ \dfrac{dy}{dt} = v \\ \dfrac{du}{dt} = \dfrac{T}{m} \cos\eta \sin\sigma + \overline{G}_X \\ \dfrac{dv}{dt} = \dfrac{T}{m} \cos\eta \cos\sigma + \overline{G}_Y \end{cases} \tag{3.57}$$

At first, let us assume constant average control angles $\overline{\eta}$, $\overline{\sigma}$, so that Eqs. (3.56) and (3.57) can be easily integrated:

$$w(t) = w_0 + \overline{G}_z\, t + f_1(t)\sin(\overline{\eta}) \tag{3.58}$$

$$z(t) = z_0 + w_0\, t + \overline{G}_z \frac{t^2}{2} + f_3(t)\sin(\overline{\eta}) \tag{3.59}$$

where the thrust integral f_3 is defined by (see value in Table 3.2):

$$f_3(t) = \int_0^t f_1(\tau)\, d\tau$$

$$u(t) = u_0 + \overline{G}_x\, t + f_1(t)\cos(\overline{\eta})\sin(\overline{\sigma}) \tag{3.60}$$

$$x(t) = x_0 + u_0\, t + \overline{G}_x \frac{t^2}{2} + f_3(t)\cos(\overline{\eta})\sin(\overline{\sigma}) \tag{3.61}$$

$$v(t) = v_0 + \overline{G}_y\, t + f_1(t)\cos(\overline{\eta})\cos(\overline{\sigma}) \tag{3.62}$$

$$y(t) = y_0 + v_0\, t + \overline{G}_y \frac{t^2}{2} + f_3(t)\cos(\overline{\eta})\cos(\overline{\sigma}) \tag{3.63}$$

The final conditions on the velocity are given: at time $t = T_{go}$ one must have: $w(T_{go}) = 0$, $u(T_{go}) = 0$, $v(T_{go}) = V_T$, so Eqs. (3.58), (3.60), (3.62) become:

$$-\,(w_0 + \overline{G}_z\, T_{go}) = f_1(T_{go})\sin(\overline{\eta}) \tag{3.64}$$

$$-\,(u_0 + \overline{G}_x\, T_{go}) = f_1(T_{go})\cos(\overline{\eta})\sin(\overline{\sigma}) \tag{3.65}$$

$$-\,(v_0 + \overline{G}_y\, T_{go}) - V_T) = f_1(T_{go})\cos(\overline{\eta})\cos(\overline{\sigma}) \tag{3.66}$$

Squaring and summing, one has a non linear equation on the time to go T_{go}:

$$f_1(T_{go})^2 = (w_0 + \overline{G}_z\, T_{go})^2 + (u_0 + \overline{G}_x\, T_{go})^2 + (v_0 + \overline{G}_y\, T_{go} - V_T)^2 \tag{3.67}$$

which can easily be solved numerically. However a real time algorithm can not rely on iterative methods of solution and a simple, albeit approximate, formula is needed. In fact in the last phase of the trajectory, the velocity is almost horizontal, so $V_0 \sim v_0$, and the gravity is almost parallel to the radial direction r_T, so $\mathbf{G} \sim G_x$. Then, if the time T_z needed to gain the final velocity is short, the time to go is well approximated by the time T_z related to the Tziolkowski law:

$$\Delta V = V_T - V_0 = V_e log(\frac{1}{1 - qT_z})$$

and the (approximate) time to go $T_{go} = T_z$ is obtained just by the difference between the target and the current velocity.

For instance, Problem 3.4 shows that the correct time to go obtained numerically is equal to $T_{go} = 3.56\,\mathrm{s}$ and the time obtained by Tziolkowsi law is $T_z = 3.58\,\mathrm{s}$.

Then Eq. (3.64) gives the average out of plane thrust angle $\overline{\eta}$:

$$\sin(\overline{\eta}) = -\frac{w_0 + \overline{G}_z\, T_{go}}{f_1(T_{go})} \tag{3.68}$$

From Eqs. (3.65), (3.66) the average in plane thrust angle $\overline{\sigma}$ follows:

$$\sin(\overline{\sigma}) = -\frac{u_0 + \overline{G}_x\, T_{go}}{f_1(T_{go})\cos(\overline{\eta})} \tag{3.69}$$

$$\cos(\overline{\sigma}) = -\frac{v_0 + \overline{G}_y\, T_{go} - V_T}{f_1(T_{go})\cos(\overline{\eta})} \tag{3.70}$$

The average control angles $\overline{\eta}, \overline{\sigma}$ take into account the final conditions on the velocity. To include the required final conditions on the position: $x(T_{go}) = r_T$, $z(T_{go}) = 0$ a small variation of the average angle is considered with a linear dependence on time, that is:

$$\eta = \overline{\eta} - k_2 + k_1\, t \tag{3.71}$$

$$\sigma = \overline{\sigma} - k_3 + k_4\, t \tag{3.72}$$

with coefficients k_i which are derived imposing the final conditions in position.

The out of plane equations (3.56) with control angle (3.71) give:

$$\dot{w} = \overline{G}_z + \frac{T}{m}\sin\overline{\eta} + \frac{T}{m}\cos\overline{\eta}(-k_2 + k_1\, t)$$

Integrating:

$$w(t) = w_0 + \overline{G}_z\, t + f_1(t)\sin\overline{\eta} - k_2\cos\overline{\eta}\, f_1(t) + k_1\cos\overline{\eta}\, f_2(t)$$

where the thrust integral f_2 is defined:

$$f_2(t) = \int_0^t \frac{T}{m}\tau\, d\tau$$

Further integration gives

$$z(t) = z_0 + w_0\, t + \overline{G}_z\,\frac{t^2}{2} + f_3(t)\sin\overline{\eta} - k_2\cos\overline{\eta}\, f_3(t) + k_1\cos\overline{\eta}\, f_4(t)$$

Table 3.2 The thrust integrals

Thrust Int.	Definition	Formula
$f_1(t)$	$\int_0^t \frac{T}{m}(\tau)\,d\tau$	$V_e log(\frac{1}{1-qt})$
$f_2(t)$	$\int_0^t \frac{T}{m}(\tau)\,\tau\,d\tau$	$-V_e\,t + \frac{1}{q} f_1(t)$
$f_3(t)$	$\int_0^t f_1(\tau)\,d\tau$	$V_e(t + (\frac{1}{q} + t) log(\frac{1}{1-qt}))$
$f_4(t)$	$\int_0^t f_2(\tau)\,d\tau$	$-V_e \frac{t^2}{2} + \frac{1}{q} f_3(t)$

The final conditions at $t = T_{go}$ and the definition of $\overline{\eta}$ give the equation of the coefficients k_1, k_2:

$$\begin{bmatrix} f_1(T_{go}) & -f_2(T_{go}) \\ f_3(T_{go}) & -f_4(T_{go}) \end{bmatrix} \begin{bmatrix} k_2 \\ k_1 \end{bmatrix} = \begin{bmatrix} 0 \\ \frac{S}{\cos(\overline{\eta})} \end{bmatrix} \tag{3.73}$$

with

$$S = z_0 + w_0 T_{go} + \overline{G}_z \frac{T_{go}^2}{2} + f_3(T_{go}) \sin(\overline{\eta})$$

The value of (3.71) is inserted in the planar equations (3.57) where

$$\sin(\sigma) = \sin(\overline{\sigma}) + \cos(\overline{\sigma})(-k_3 + k_4\,t)$$

$$\cos(\sigma) = \cos(\overline{\sigma}) - \sin(\overline{\sigma})(-k_3 + k_4\,t)$$

The values k_3 and k_4 can be determined using the same procedure of the out of plane case and imposing the final conditions $x(T_{go}) = r_T$, $u(T_{go}) = 0$.

The above formulae can be applied at each guidance step updating the current state (obtained for instance by the navigation system) and repeating the procedure with updated time to go, which is decreasing since the variation of velocity $DV = V_T - V_0$ is decreasing during the approach to the orbit injection point and so it is the time T_z approximating T_{go}.

The above formulae depend on the thrust integrals defined in Table 3.2 (recall that the thrust force is approximated by $T = g\,Isp\,m_d = V_e\,m_d$ and $m = m_0 - m_d\,t$, $q = \frac{m_d}{m_0}$):

Problems

3.1 Determine the numerical trajectory of Scout G with payload mass of 150 kg until injection into circular orbit of altitude $h_f = 650$ km and inclination $i = 40$ deg, RAAN $\Omega = 250\,deg$ by time optimal guidance of the fourth stage.

3.2 Estimate the maximum payload that can be set into the circular orbit of altitude $h = 650\,\text{km}$, $i = 40\,\text{deg}$ and $RAAN = 250\,\text{deg}$

3.3 Avum is a liquid fueled stage of the Vega launcher, capable of multiple restarts. Suppose that a first mission deployment has been achieved at circular orbit of 500 km of altitude. A second spacecraft is to be deployed at 800 km of altitude in circular co-planar orbit. The launcher mass is 700 kg, the mass rate is 0.85 kg/s and thrust is 2420 N. Apply the Q-guidance in position and velocity to achieve the target orbit and compute the propellant mass used.

3.4 Let the last stage Altair of the Scout launcher have orbit parameters $\Omega = 308\,\text{deg}$, $i = 39.95\,\text{deg}$, $\omega = 179\,\text{deg}$, $\theta = 180\,\text{deg}$, apogee altitude $h_a = 690$ km and perigee altitude $h_p = 165$ km. The structural mass is $m_s = 218\,\text{kg}$, the payload mass is $m_{pay} = 150\,\text{kg}$ the propellant mass is $m_p = 286\,\text{kg}$, the specific impulse is $Isp = 218\,\text{s}$ and the burn out time $t_b = 34\,\text{s}$. Implement a tangent linear law to achieve the circular orbit of altitude 700 km, RAAN $\Omega_T = 308$ deg and inclination $i_T = 40\,\text{deg}$.

Solutions

Problems of This Chapter

3.1 The Scout trajectory with guidance parameters $t_p = 17.8\,\text{s}$ and $t_c = 574\,\text{s}$ is considered with the other guidance parameters fixed to $kick = 0.5\,\text{deg}$, $t_v = 6\,\text{s}$, $\psi_A = 132.7\,\text{deg}$. The state at the end of the coasting time is:

$$h_0 = 647\,\text{km}\ ,\ V_0 = 6378\,\text{m/s}\ ,\ \gamma_0 = 2.33\,\text{deg}$$

After the projection on the TPI frame, the ECI position coordinates after coasting are transformed to:

$$x_0 = 7025\,\text{km}\ ,\ y_0 = 0.0\,\text{km}\ ,\ z_0 = 0.0\,\text{km}$$

and the ECI velocity has TPI coordinates:

$$vx_0 = 0.258\,\text{km/s}\ ,\ vy_0 = 6.381\,\text{km/s}\ ,\ vz_0 = 0.0\,\text{km/s}$$

The minimum time guidance is implemented by guessing the initial and final thrust angle:

$$\theta_0 = 2\,\text{deg}\ ,\ \theta_f = 0\,\text{deg}$$

The algorithm quickly converges to the control angle direction of Fig. 3.18, where the altitude and velocities along the two axis are also plotted. It is apparent that:

Table 3.3 Data of flight, first three stages, payload 250 kg

time (s)	Event	Altitude (km)	Rel. Vel (m/s)	Rel. fpa (deg)
0	Lift off	0.0	0.0	89.5
6	End of vertical arc	0.2	123	89.3
40	Pitch over	11	763	40
80	First stage burn out	39	1540	46.0
119	Second stage burn out	74	3757	32.6
166	Third stage burn out	141	6378	27.6
698	End of coasting	651	5640	2.0

(i) the final orbit is achieved: altitude is 650 km, the velocity is 7.5 k/s with no radial component (orbital velocity)

(ii) the minimum time is only 20 s, which is less than the fourth stage burn out.
3.2 Scout G launcher is assumed to have a payload mass of 250 kg. This is a little less than the previous payload mass of 150 kg and the propellant saved by the minimum time guidance. The target orbit is a circular orbit of altitude $h = 650$ km, inclination $i = 40$ deg, RAAN $\Omega = 250$ deg.

The following guidance parameters are selected:

- azimuth angle $\psi_A = 132.7°$, launch time: 22nd October 2026 0.0:0.0:0.0
- duration of vertical arc $t_v = 6$ s. Kick angle $\chi = 0.5$ deg
- duration of pitch manoeuvre $t_p = 16.7$ s
- duration of coasting time $t_c = 524$ s

Table 3.3 shows the main events occurring during the flight up to coasting. The variation of RAAN and inclination is in Fig. 3.26, so the out of plane orbit parameters are met. The velocity at the end of coasting is very close to the velocity derived by analytic formula of Sect. 3.3. The minimum time algorithm is implemented after the coasting. The output is injection into orbit with minimum time equal to the burn out time of the Scout fourth stage $t_{b4} = 34$ s. Figure 3.27 shows that the target altitude of 650 km is reached. The velocity reached is equal to the orbit velocity with no radial component. The guidance angle increases from the −45° to 14°. As the minimum thrust time coincides with the burn out time, the payload mass of 250 kg is the maximum mass that can be inserted into the circular orbit of altitude 650 km.

3.3 The guidance in position is implemented with the initial state shown by the red circle in the plot of the curve of necessary velocities to reach the altitude 800 km from 500 km and transfer angle $\phi = 80$ deg, see Fig. 3.28.

The velocity to be gained is computed updating at each second the current position and the curve of necessary velocities: after 41 s the curve is reached and the coasting arch starts. The radial and tangential velocities during the guidance in position

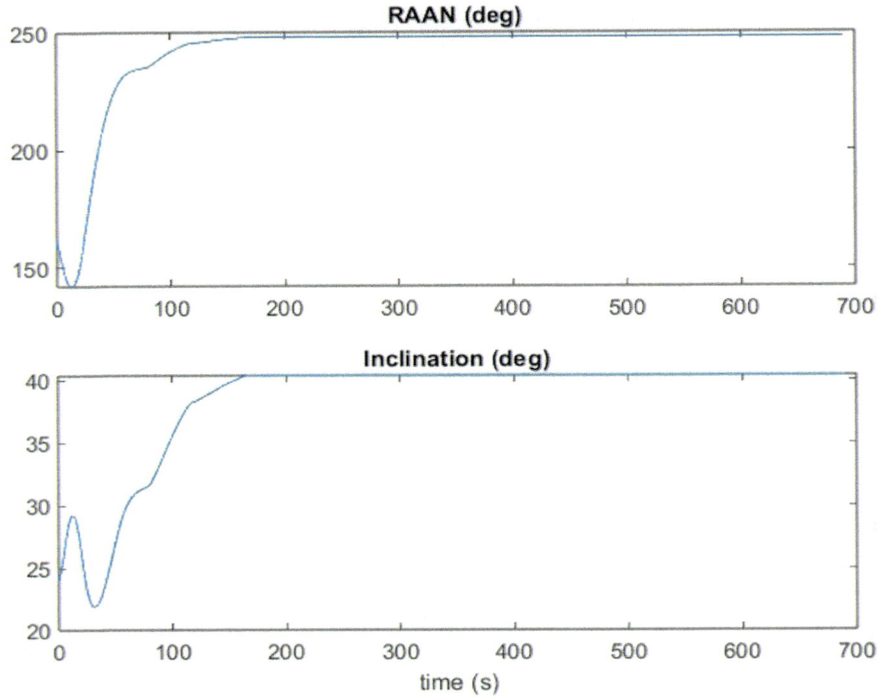

Fig. 3.26 RAAN and inclination angles after the first three stages of the Scout launcher, including coasting. Payload 250 kg

are in Fig. 3.29. The guidance in velocity starts at the end of coasting of duration 1410 s, $\phi = 0.45$ deg, and state conditions $x_0 = 7167.3$ km, $y_0 = -56.30$ km, $u_0 = 0.132$ km/s, $v_0 = 7.330$ km/s in TPI coordinates. The target velocity and position are reached after 52 s, see the radial-tangential velocities in Fig. 3.30. Figure 3.31 shows the variation of energy during the transfer. The boost phase lasts 93 s and the propellant consumed is 79 kg.

3.4 The orbital conditions at the Altair ignition correspond to the following ECI coordinates:

$$X_0 = 4362\,\text{km}\,, \;\; Y_0 = -5607\,\text{km}\,, \;\; Z_0 = -79\,\text{km}$$

$$VX_0 = 4.51\,\text{km/s}\,, \;\; VY_0 = 3.39\,\text{km/s}\,, \;\; Z_0 = 4.72\,\text{km/s}$$

These state coordinates are rotated to the target plane, giving the following result:

$$x_0 = 7067\,\text{km}\;, \;\; y_0 = -128\,\text{km}\;, \;\; z_0 = 0.107\,\text{km}$$

$$u_0 = 0.133\,\text{km/s}\;, \;\; v_0 = 7.362\,\text{km/s}\;, \;\; w_0 = -0.006\,\text{km/s}$$

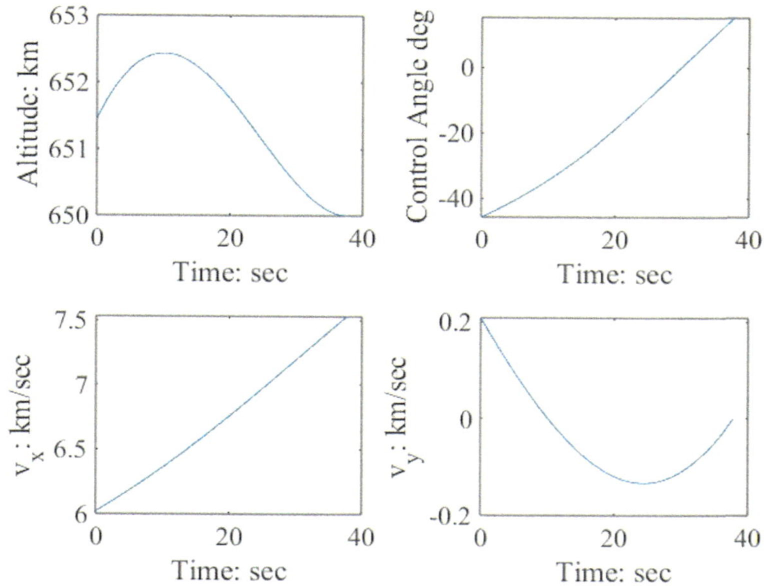

Fig. 3.27 The insertion into circular orbit of altitude 650 km by time optimal guidance, payload 250 kg

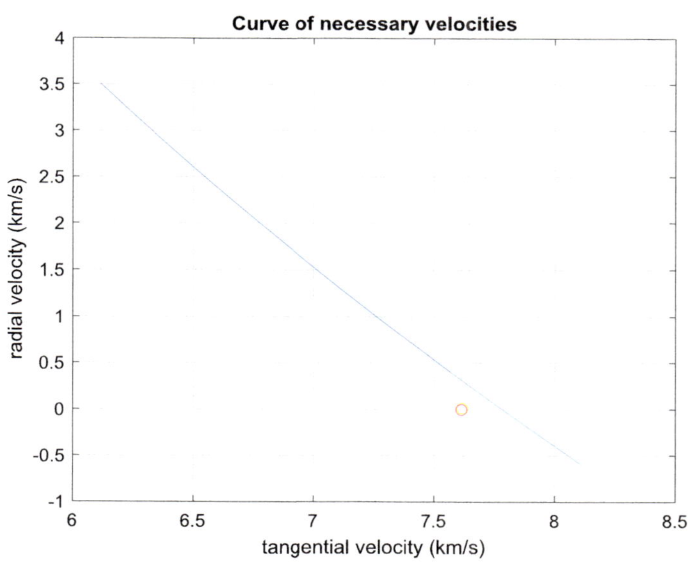

Fig. 3.28 The curve of necessary velocities at the initial time of Problem 3.3

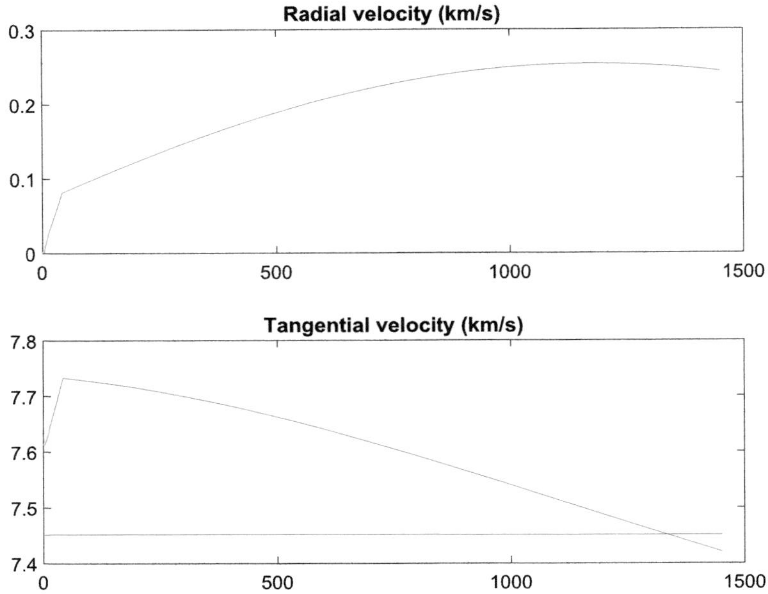

Fig. 3.29 The radial and tangential velocities during the guidance in position and coasting arch

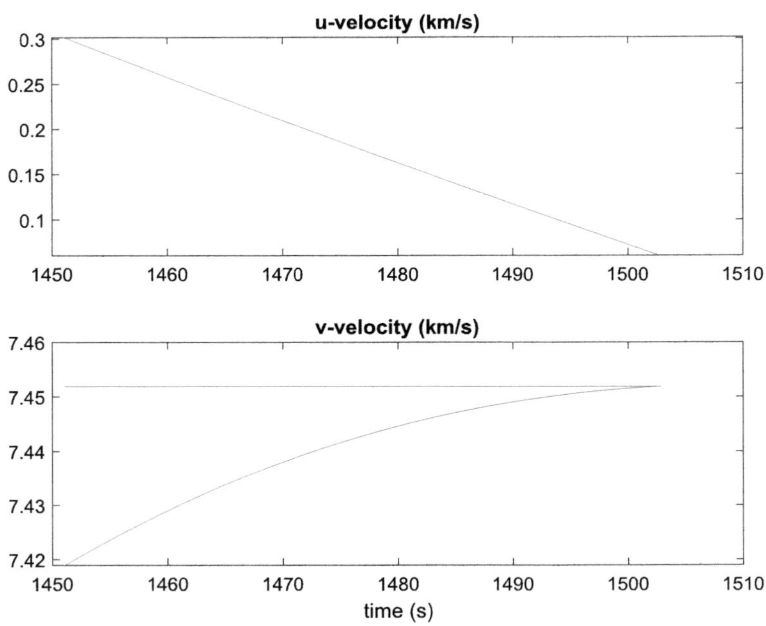

Fig. 3.30 The radial and tangential velocities during the guidance in velocity

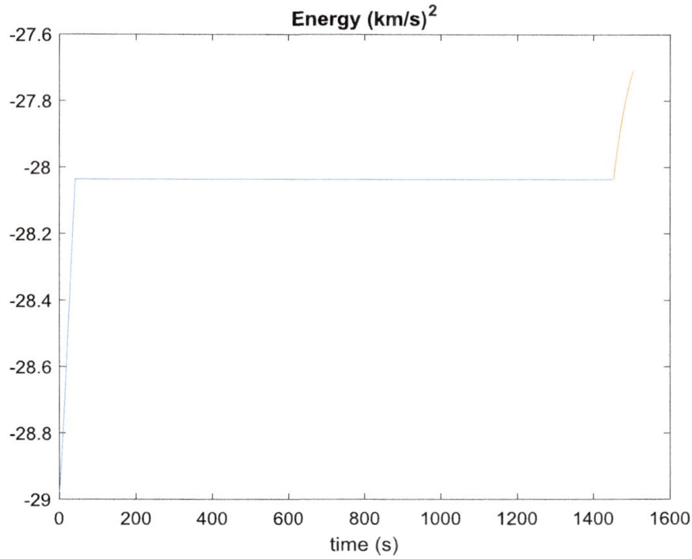

Fig. 3.31 Avum energy during the transfer trajectory

The angle between the target plane nodal line and the projection of the radius vector on the target orbit plane is $\phi_0 = -1$ deg.

The time to go obtained by formula (3.67) gives $T_{go} = 3.56$ s. The required $\Delta V = V_T - V_0$ is equal to 140 m/s, then the time needed to achieve this variation of velocity is, according to the Tziolkowski formula: $T_z = 3.51$ s and this is taken here as the time to go.

The flight path angle corresponding to the ECI initial state is $\gamma_0 = 0.4$ deg, then the transfer angle defined by the formula (3.53) is equal to: $\Phi = -0.66$ deg. As a result the initial state of the launcher with respect to the Target Orbital frame has coordinates:

$$x_{n_0} = 7068\,\text{km}\ ,\ y_{n_0} = 0\text{km/s}\,\text{km}\ ,\ z_{n_0} = 0\,\text{km}$$

$$u_{n_0} = 0.005\,\text{km/s}\ ,\ v_{n_0} = 7.363\text{km/s}\,\text{km}\ ,\ w_{n_0} = -0.006\,\text{km/s}$$

According to the formulas (3.68), (3.69), (3.70), it is found:

$$\overline{\eta} = 2.58\,\text{deg}\ ,\ \overline{\sigma} = -8.38\,\text{deg}$$

Figure 3.32 shows that the out of plane dynamics is annihilated by the Altair burn.

Figure 3.33 shows that the average control angles $\overline{\eta}$, $\overline{\sigma}$ are suitable to achieve the required orbital velocity within the time to go. The Figs. 3.34 and 3.35 show that the final conditions in the position are almost fulfilled even without the corrections to the average control angles of formulas (3.71), (3.72).

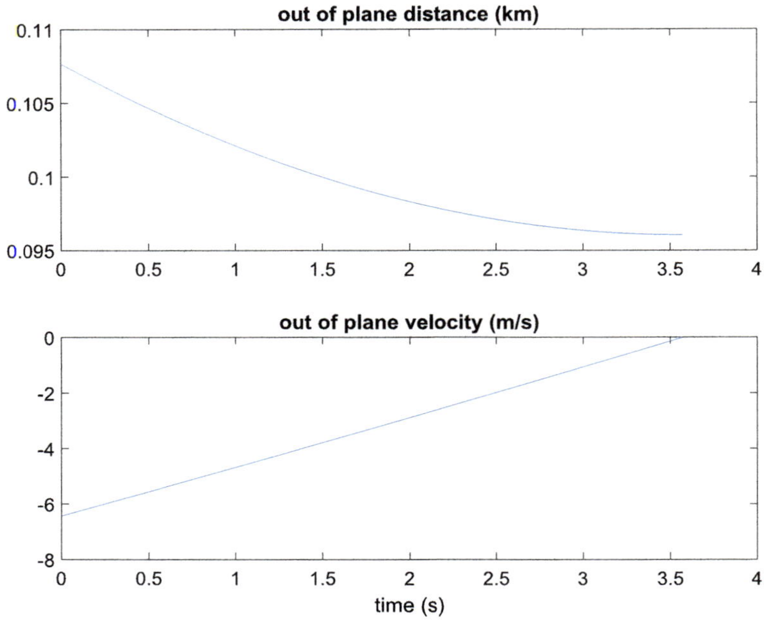

Fig. 3.32 The out of plane velocity and position during the real time tangent linear guidance

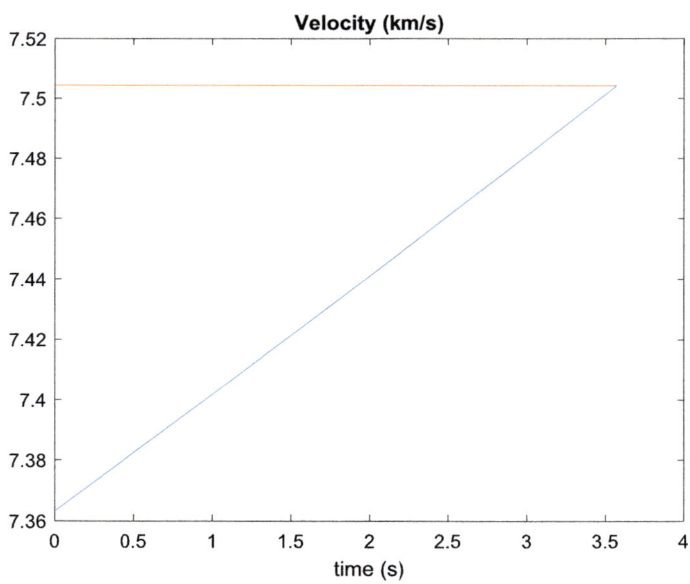

Fig. 3.33 The total velocity gained during the real time tangent linear guidance and the target velocity

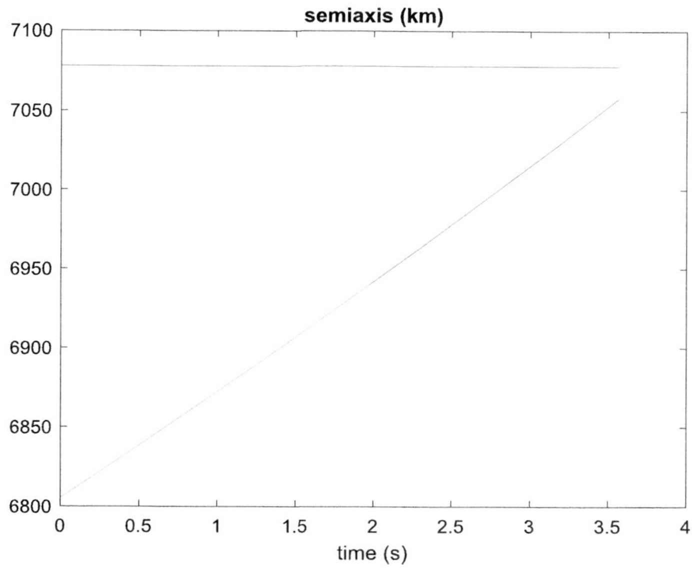

Fig. 3.34 The semiaxis variation during the real time tangent linear guidance and the target semiaxis

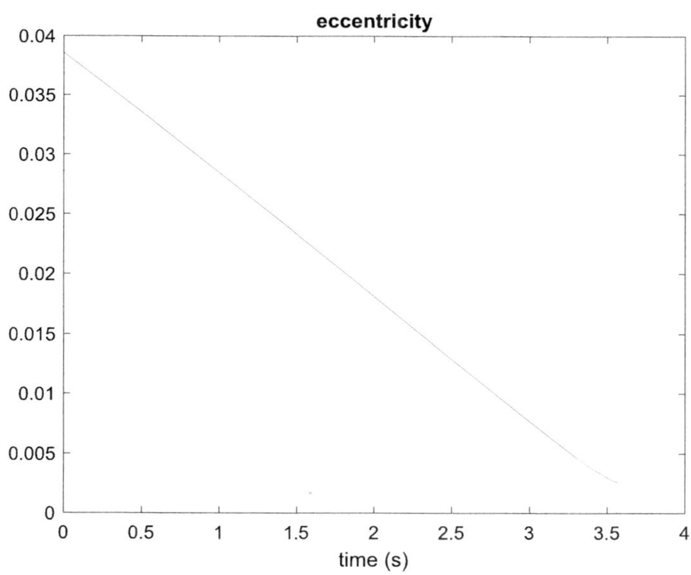

Fig. 3.35 The eccentricity variation during the real time tangent linear guidance (the target eccentricity is zero)

References

1. Marec, J.: Optimal Space Trajectories. Elsevier, New York (1979)
2. Bryson, A.E., Ho, Y.: Applied Optimal Control. Hemisphere, New York (1975)
3. Prussing, J.: Optimal Spacecraft Trajectories. Oxford University Press (2018)
4. Miele A.: Flight Mechanics: The Theory of Flight Paths. Addison-Wesley Publishing (1962)
5. Lawden, D.: Optimal Trajectories for Space Navigation. Butterworth, London (1963)
6. Rimrott, F.: Introductory Orbit Dynamics. Vieweg+Teubner Verlag (1989)
7. Pontani, M., Pianalto, A., Carletta S., Teofilatto, P.: Near Optimal guidance and pulse modulated reduced attitude control for orbit injection. In: Fasano, G., Pintér, J.D. (eds.), Modeling and Optimization in Space Engineering. Springer Optimization and Its Applications, vol. 200. Springer (2023)
8. Battin, R.H.: An Introduction to the Mathematics and Methods of Astrodynamics. AIAA Education Series. American Institute of Aeronautics and Astronautics, New York (1987)
9. Rinalducci A.: A guide system for a multi stage launcher. Special Master Degree Thesis, Scuola di Ingegneria Aerospaziale di Roma (1997)
10. Townsend, G.E., Abbot, A.S., Palmer, R.R.: Guidance, flight mechanics and trajectory optimisation, vol. III– boost guidance equations. Nasa Report SID 65–1200-8 (1968)

Chapter 4
Reentry Trajectories

Abstract Reentry trajectories have been studied since the beginning of the space activity. In 1961 Yuri Gagarin performed a safe re-entry into the atmosphere from a Low Earth Orbit and soon after the Apollo missions began reentry from the Moon. There are two major issues in re-entry missions: the peak deceleration produced by the drag force and the peak heat flux produced by the conversion of the kinetic energy loss into thermal energy. The heat and deceleration peaks constraints are satisfied by trajectories that fly within a specified region called the entry corridor. In the case of ballistic reentry (negligible lift) the entry corridor can be determined by analytical formulae derived here that patch the Kepler solution (no drag) with the Allan-Egger formula (no gravity). To increase the entry corridor and control the trajectory to the desired landing point, a small aerodynamic efficiency is required. The three-dimensional equations for re-entry with lift are then analysed together with algorithms for correcting the point of impact.

4.1 Introduction

Reentry has been studied since the early days of the space exploration: since the Yuri Gagaring mission (1961) it was necessary to design a safe reentry trajectory towards the Earth. The reentry capsule of that mission was the Vostok capsule which had a blunt body shape. This is the appropriate configuration to reduce the high temperatures occurring during a reentry trajectory. Other important applications of the re-entry dynamics include the trajectories of the exhausted stages of a launcher and the heliocentric trajectories of asteroids that, passing through the Earth atmosphere, can approach or impact the Earth surface. In all the above examples the lift component of the aerodynamic force is rather small and this is the condition of ballistic reentry. However some lift control effect is used in some reentry vehicles: these can have from a moderate aerodynamic efficiency up to the high lift of winged reentry systems capable of landing.

The major drivers in the reentry are the structural and thermal loads due to the drag effect of the air resistance on the reentry capsule; in particular the reentry trajectory must be designed to satisfy the limits on the peak of deceleration and on the peak of

P. Teofilatto, *Trajectories of Launch and Reentry Systems*, Springer Aerospace Technology, https://doi.org/10.1007/978-3-031-75760-0_4

temperature [1, 2]. In fact the trajectories allowing a safe flight into the atmosphere must be inside a corridor (reentry corridor), see Fig. 4.1. The reentry corridor is a region bounded by two curves: the external boundary curve is called "undershoot " boundary, the other is called "overshoot" boundary. Outside the reentry corridor on the undershoot side (for instance trajectory "B" in the Figure) the ΔV maneuvre to reduce the orbital velocity and to reentry is too low, and the trajectory will not return to Earth but it will be a new keplerian orbit with a reduced semiaxis, Fig. 4.2. Outside the reentry corridor on the overshoot side the decrement of velocity ΔV to return to Earth is too big: the thermal or the deceleration constraints are violated and the reentry caspule is destroyed (trajectory "A" in Figs. 4.1 and 4.2, right).

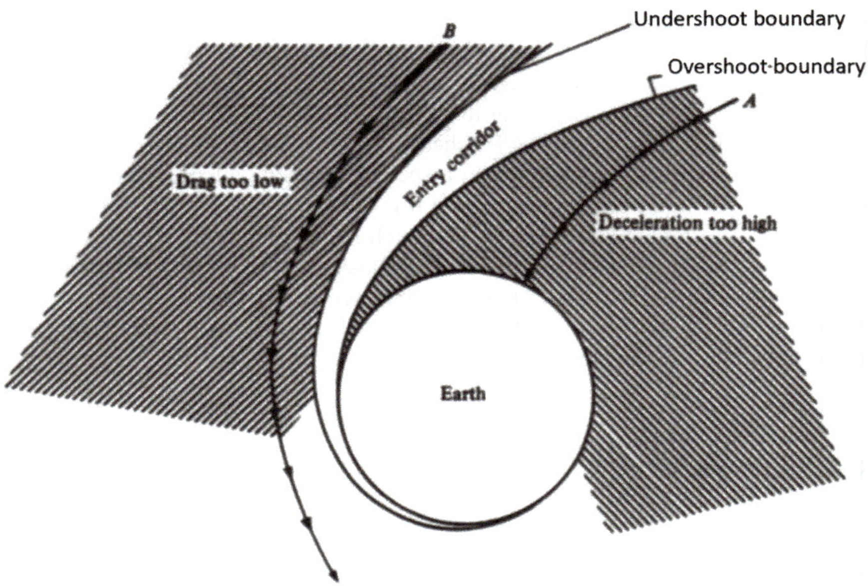

Fig. 4.1 Reentry corridor

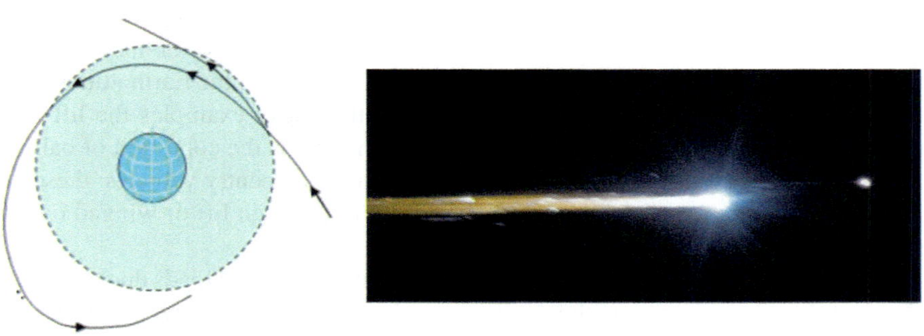

Fig. 4.2 Undershoot trajectory (left) and capsule in overshoot region (right)

Apollo capsule

Soyouz capsule

Fig. 4.3 The Apollo and Soyuz capsules

The entry corridor can be enlarged by the lift which can substain the reentry vehicle avoiding too steep trajectories. The lift can also be used to change the attitude of the vehicle and to control the trajectory. Namely, to reduce the thermal stresses and at the same time have some flexibility in trajectory control, reentry capsules are blunt shaped with a moderate aerodynamic efficiency, around 0.3, see Figs. 4.3 and 4.4. Some lift is essential for return at the high velocities of interplanetary or lunar trajectories [3]. Figure 4.5 shows the reentry strategy of the Apollo missions: a lift controlled manoeuvre shortly after the first entry point allows to follow a skip trajectory to leave the atmosphere and to return inside it at a lower velocity and flight path angle [4]. A direct reentry from lunar orbit will make the entry corridor too small to be safe. Table 4.1 shows the main entry data of different Apollo missions. Note that the maximum load is around 6-7g, rather high values due to the hyperbolic reentry velocities. The table also gives the total heat integrated over time. In fact such trajectories have a long duration within the atmosphere and it is important to compute the total heat accumulated together with its peak. On the other hand, it should be noted that the Mercury missions, returning from Earth orbits, were able to perform direct reentry trajectories. Direct reentry is also used by unmanned capsules returning from Earth orbits. Generally these capsules can withstand higher peak loads, say around 15 g, with low aerodynamic efficiency and limited control capability.

Space reentry vehicles endowed with relevant aerodynamic surfaces, allow also a controlled landing [5]. Figure 4.6 and table of Fig. 4.7 show some of these systems, the most famous of which is the Shuttle. These systems have rather high aerodynamic efficiency, taking into account also the reduced drag coefficient compared to the blunt shaped capsules. Figure 4.8 shows the variation of the aerodynamic drag coefficient C_D for different reentry blunt shaped capsules [6]. Note the large variation of C_D from subsonic Mach values to supersonic and hypersonic Mach values in the case of the blunt capsules of Fig. 4.8, see [6–8]. On the other hand the winged reentry vehicles

Symbol	Quantity	APOLLO	SOYUZ
A_{ref}	reference area $[m^2]$	12.02	3.80
L/D	aerodynamic performance	0.30	0.26
C_L	lift coefficient	0.374	0.349
C_D	drag coefficient	1.247	1.341
m	total vehicle mass [kg]	5,470.0	2,400.0
V_e	flight velocity at entry [m/s]	7,670.0	7,900.0
γ_e	flight path angle at entry [°]	-0.75, -1.50, -3.50	-1.50
R_N	radius at stagnation point [m]	4.694	2.235

Fig. 4.4 The Apollo and Soyuz main data

Table 4.1 Entry data of Apollo missions

	Apollo 8	Apollo 10	Apollo 11	Apollo 12	Apollo 13	Apollo 14	Apollo 16
Lift to Drag	0.30	0.30	0.30	0.31	0.29	0.28	0.28
Velocity (m/s)	11040	11068	11032	11008	11037	11025	11032
Flight path angle (deg)	–6.5	6.5	–6.5	–6.5	–6.3	–6.4	–6.5
a_{max} (g)	6.8	6.8	6.5	6.6	5.6	6.7	6.2
q_{max} (KW/m^2)	3360	3360	3246	3235	3076	3518	3927
Q (KJ/m^2)	296689	292010	300570	297640	291810	307710	317110

have a relevant C_D variation only around the transonic regime (Fig. 4.9), that is in a limited arc of the reentry trajectory: this is a typical behaviour of slender-winged bodies.

4.2 Ballistic Reentry

Ballistic reentry is studied using the analytic formulae due to Allan and Egger [9] which hold true at low altitudes, where the drag acceleration overcomes the gravity. With these formulas it is possible to determine the peak of deceleration and the the

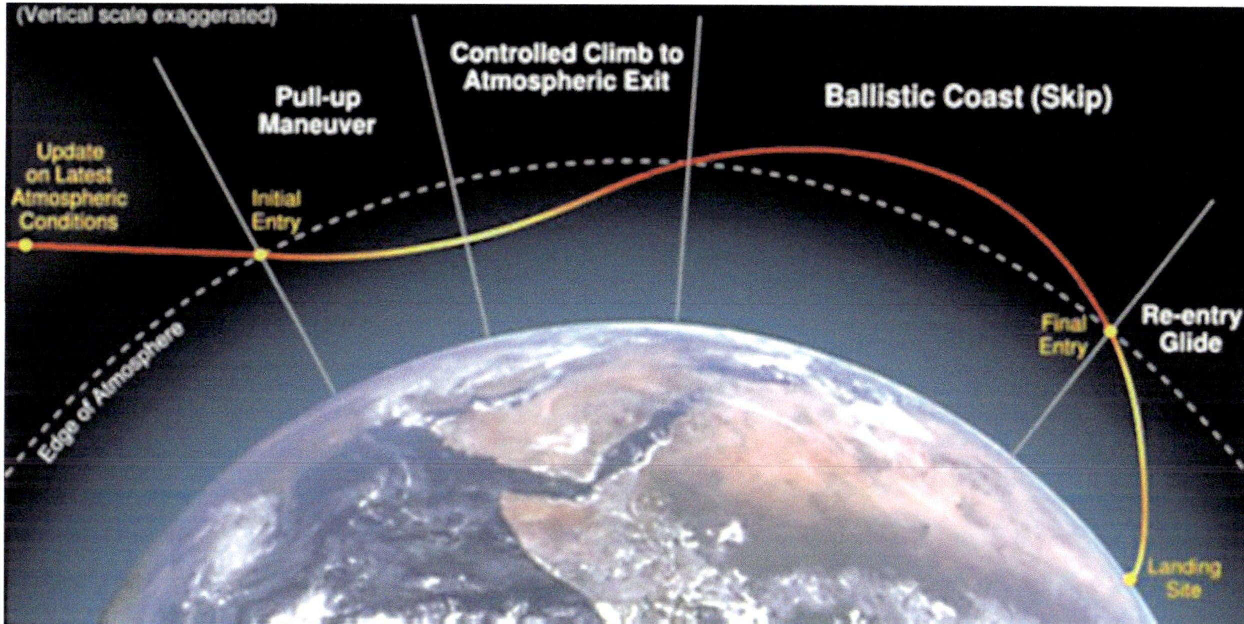

. Illustration of the SKIP re-entry (red line) stated by Chris Kraft for Apollo and now considered absolutely essential for the Orion lunar return.

Fig. 4.5 The skip reentry of Apollo mission. Image credit: Clem Tillier; Earth graphic based on NASA image of Earth seen from Apollo 17, CC BY-SA 2.5

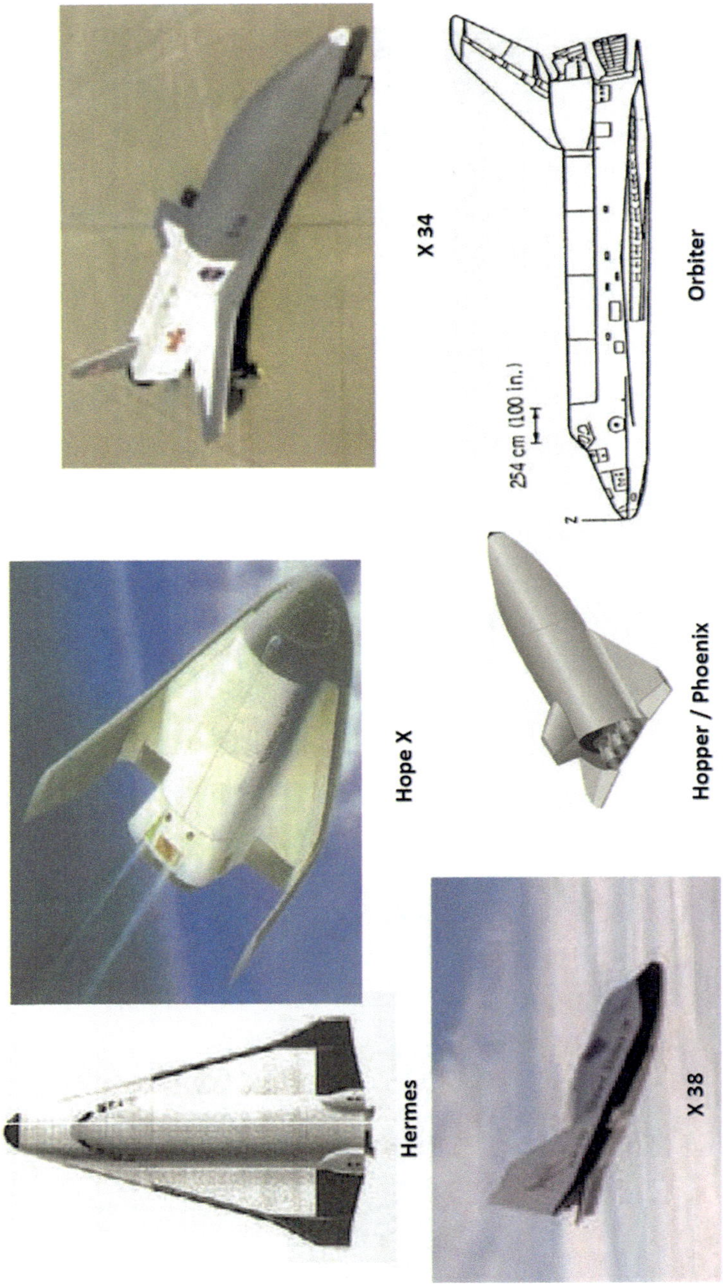

Fig. 4.6 Winged reentry vehicles

Vehicle	L_{ref} [m]	A_{ref} [m^2]	Wing span [m]	Sweep angle [°]	Strake angle [°]	x_{cog}/L_{ref} [–]
Space Shuttle Orbiter	32.774	249.91	23.790	45.0	81.0	0.650
HERMES	15.500	73.000	9.375	74.0	–	0.600
HOPE-X	15.249	50.220	9.670	55.0	75.0	0.635
X-34	17.678	33.213	8.534	45.0	80.0	0.600
X-38	8.410	21.670	3.658	–	–	0.570
HOPPER	43.650	545.120	27.200	55.0	–	0.680
PHOENIX	6.900	15.14	3.84	62.5	–	0.700

Fig. 4.7 Some data of winged reentry vehicles, from [5]

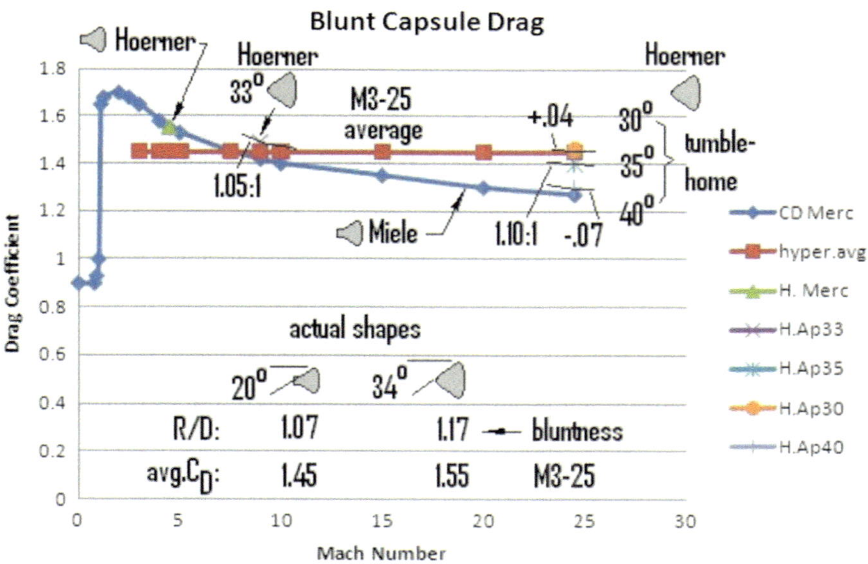

Fig. 4.8 Drag coefficient of blunted reentry capsule, image credit: Prof. G. W. Johnson [6]

peak of temperature as a function of the state of the capsule at the "reentry point". It will be possible to demonstrate that the maximum deceleration depends only on the state of the capsule and not on its shape and mass, whereas the maximum thermal flux depends on the mass and geometry of the capsule, in particular on the radius of curvature of the capsule surface at the point of flow stagnation. The larger the radius

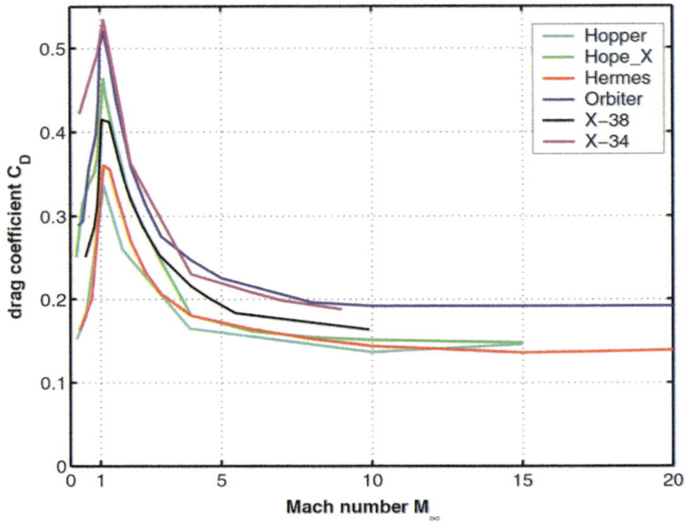

Fig. 4.9 Drag coefficient of winged reentry vehicles [5]

of curvature the smaller the incoming thermal flux; then reentry capsules are blunt bodies. The Allan and Egger formulas, and the related computation of the reentry parameters, neglect the gravity effect and rely on a vague definition of a reentry point. In fact the reentry trajectory is a transition from a keplerian to an atmospheric orbit and a formula is proposed that is uniformly valid along the entire flight. This improved formula makes it possible to relate the maximum deceleration and thermal flux to the initial Keplerian conditions. The correct velocity variation for a safe reentry can then be evaluated at the beginning of the reentry mission.

The relevant frames in this analysis are orbital frame $(\hat{r}, \hat{\theta}, \hat{h})$ and the velocity frame of Fig. 4.10:

It has to be recalled that the aerodynamic force depends on the relative velocity

$$\mathbf{V}_R = \mathbf{V} - (\boldsymbol{\omega}_E \wedge \mathbf{r} + \mathbf{w}_{loc})$$

However the absolute velocity $\mathbf{V}$ is about 7000–11000 m/s while the wind velocity is less than 500 m/s, then it is reasonable to neglect the wind velocity and to consider only the absolute velocity in the formulas of the lift and drag. Furthermore the so called ballistic coefficient B and the aerodynamic efficiency E are introduced:

$$B = \frac{1}{2}\frac{C_D S}{m}$$

$$E = \frac{C_L}{C_D}$$

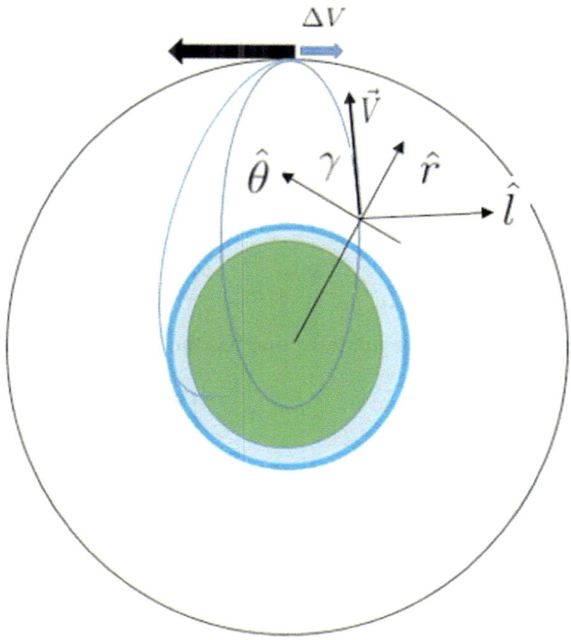

Fig. 4.10 Orbital and velocity frames

and the following formulas will be considered:

- Drag: $-\rho B V^2 \hat{V}$
- Lift: $\rho B E\, V^2 \hat{l}$.

Due to the very high regime of velocity, we can regard the aerodynamic coefficients as constant with respect to the Mach number, see Fig. 4.8. The dynamic equation of the reentry is:

$$[\frac{d\mathbf{V}}{dt} = -\mathbf{g} + \frac{\mathbf{D}}{m} + \frac{\mathbf{L}}{m} = -\frac{\mu}{r^2}\hat{r} - \rho B V^2 \hat{V} + \rho B E V^2 \hat{l}]$$

and, taking into account Fig. 4.10, $\hat{r} = \sin\gamma\,\hat{v} + \cos\gamma\hat{l}$:

$$\frac{d\mathbf{V}}{dt} = -\rho V^2 B\,\hat{v} - \frac{\mu}{r^2}\cos\gamma\,\hat{l} - \frac{\mu}{r^2}\sin\gamma\,\hat{v} + \rho V^2 B\,E\,\hat{l}$$

Also the left hand side of the above equation must be written in the non inertial frame $(\hat{l}, \hat{V}, \hat{h})$ which rotates with respect to the inertial frame by the angular velocity $(\dot{\theta} - \dot{\gamma})\hat{h}$, hence:

$$\frac{d\mathbf{V}}{dt} = \dot{V}\hat{v} + \boldsymbol{\omega} \wedge \mathbf{V} = \dot{V}\hat{v} + (\dot{\theta} - \dot{\gamma})\hat{h} \wedge V\hat{v} = \dot{V}\,\hat{v} + (\dot{\gamma} - \dot{\theta})V\,\hat{l}$$

Therefore the scalar equations of the (planar) reentry are:

$$\begin{cases} \hat{v}: & \dot{V} = -\frac{\mu}{r^2}\sin\gamma - \rho B V^2 \\ \hat{l}: & \dot{\gamma} = \dot{\theta} - \frac{\mu}{r^2 V}\cos\gamma + \rho B\, E\, V \end{cases}$$

These are paired with the kinematic equations which are obtained writing the velocity vector with respect to the orbital frame $(\hat{r}, \hat{\theta}, \hat{h})$:

$$\mathbf{V} = \frac{d\mathbf{r}}{dt} = \dot{r}\,\hat{r} + \dot{\theta}\hat{h} \wedge \mathbf{r} = \dot{r}\,\hat{r} + r\dot{\theta}\,\hat{\theta}$$

Using the above relations one has the kinematic equations:

$$\begin{cases} \dot{r} = V\sin\gamma \\ \dot{\theta} = \frac{V}{r}\cos\gamma \end{cases}$$

and finally the planar equations of the reentry:

$$\dot{V} = -\frac{\mu}{r^2}\sin\gamma - \rho B V^2 \tag{4.1}$$

$$\dot{\gamma} = \left(\frac{V}{r} - \frac{\mu}{r^2 V}\right)\cos\gamma + \rho B\, E\, V \tag{4.2}$$

$$\dot{r} = \dot{h} = V\sin\gamma \tag{4.3}$$

$$\dot{\theta} = \frac{V}{r}\cos\gamma \tag{4.4}$$

The Allan-Egger Solution

Generally capsules for ballistic reentry have aerodynamic efficiency lower than $E = 0.1$, then it is reasonable to approximate $E \simeq 0$, then the Eq. 4.2 for the flight path angle becomes:

$$\dot{\gamma} = \left(\frac{V}{r} - \frac{\mu}{r^2 V}\right)\cos\gamma$$

A "flat Earth" model will be assumed: the flat Earth hypothesis can implemented in the equations letting the Earth radius $R_\oplus$ go to infinity. Then the reentry equations become:

$$\dot{V} = -\frac{gR_\oplus^2}{R_\oplus^2}\frac{1}{\left(1 + \frac{h}{R_\oplus}\right)^2}\sin\gamma - \rho B V^2$$

$$\dot{\gamma} = \left(\frac{V}{R_\oplus + h} - \frac{gR_\oplus^2}{R_\oplus^2}\frac{1}{\left(1 + \frac{h}{R_\oplus}\right)^2}\frac{1}{V}\right)\cos\gamma$$

and for $R_\oplus \to \infty$ one has the equations of ballistic reentry in the flat Earth hypothesis:

$$\dot{V} = -g \sin\gamma - \rho B V^2 \tag{4.5}$$

$$\dot{\gamma} = -\frac{g}{V} \cos\gamma \tag{4.6}$$

$$\dot{h} = V \sin\gamma \tag{4.7}$$

Note that during the reentry the flight path angle γ is negative, so the gravity term accelerates the capsule, while the drag always slows it down. At high altitudes the gravity term dominates then the drag, reducing the semiaxis of the initial orbit, produces an increasing velocity (the "paradox of astrodynamics"), but as the capsule approaches the Earth, the air density ρ increases exponentially and at a certain point the drag term becomes dominant. Figure 4.11 shows the velocity behaviour for a typical ballistic reentry. The capsule starts from circular orbit at 300 km of altitude, and after a ΔV for the reentry, the flight path angle is equal to –5 deg when the capsule is at altitude 280 km. Note the small but visible increase in velocity up to an altitude of about 85 km. From this point on, the drag becomes dominant and it is able to decelerate the capsule.[1]

When this happens we fix the state conditions at the "entry point": V_e, γ_e, r_e and consider the following equations of the motion:

$$\dot{V} = -\rho B V^2 \tag{4.8}$$

$$\dot{\gamma} = 0 \tag{4.9}$$

$$\dot{h} = V \sin\gamma \tag{4.10}$$

where the gravity has been neglected.

Note that the approximation $\dot{\gamma} = 0$ implies that the reentry trajectory is a straight line with slope $\gamma = \gamma_e$ (*entry angle*). Therefore

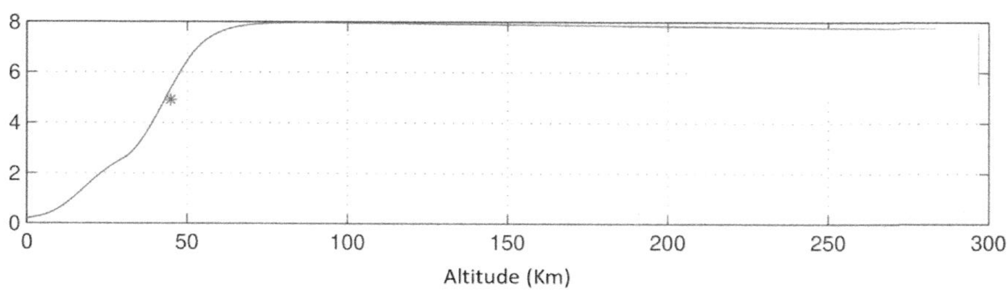

Fig. 4.11 The velocity (in km/s) during a ballistic reentry trajectory as function of the altitude

[1] The change in the slope of the velocity decay around 30 Km of altitude is due to the change in the drag coefficient with the transition to the continuous flow regime.

$$\dot{V} = -\rho B V^2, \qquad \dot{h} = V \sin\gamma_e \qquad (\gamma_e < 0)$$

Since

$$\dot{V} = \frac{dV}{dt} = \frac{dV}{dh}\dot{h}$$

one has:

$$\frac{dV}{dh} = \frac{\dot{V}}{\dot{h}} = -\frac{\rho B}{\sin\gamma_e} V = -\frac{\rho_0 B}{\sin\gamma_e} e^{-\beta h} V$$

The ballistic coefficient B is considered as constant in the hypersonic regime and it is possible to integrate the equation of motion

$$\frac{dV}{V} = -\frac{\rho_0 B}{\sin\gamma_e} e^{-\beta h} dh \quad \Rightarrow \quad \int_{V_e}^{V} \frac{dV}{V} = \int_{h_e}^{h} \left(-\frac{\rho_0 B}{\sin\gamma_e}\right) e^{-\beta h} dh \quad \Rightarrow$$

$$\ln\frac{V}{V_e} = \left(-\frac{\rho_0 B}{\sin\gamma_e}\right)\left[-\frac{1}{\beta} e^{-\beta h}\right]_{h_e}^{h} = \left(-\frac{\rho_0 B}{\sin\gamma_e}\right) \frac{e^{-\beta h} - e^{-\beta h_e}}{-\beta} \quad \Rightarrow$$

$$V = V_e \exp\left\{\frac{\rho_0 B}{\beta \sin\gamma_e}\left(e^{-\beta h} - e^{-\beta h_e}\right)\right\} \simeq V_e \exp\left\{\frac{\rho_0 B}{\beta \sin\gamma_e} e^{-\beta h}\right\} \tag{4.11}$$

The above formula for the velocity during a reentry trajectory is called double exponential formula [9].

Using the above formula it is possible to compute the altitude where the peak of deceleration occurs and the value of the peak of deceleration. In the case of reentry with a crew it is necessary to limit the deceleration to 4 g / 6 g; in the case of unmanned reentry capsule, a typical limit is around 15 g.

By the double exponential formula the drag deceleration $\dot{V}$ is equal to:

$$\dot{V} = -\rho B V^2 = -\rho_0 e^{-\beta h} B V^2 = -\rho_0 e^{-\beta h} B V_e^2 \exp\left\{2 \frac{\rho_0 B}{\beta \sin\gamma_e} e^{-\beta h}\right\}$$

that is:

$$\dot{V} = -\rho_0 B V_e^2 \exp\left\{\frac{2\rho_0 B}{\beta \sin\gamma_e} e^{-\beta h} - \beta h\right\} \tag{4.12}$$

At the peak of the deceleration $\frac{d\dot{V}}{dh} = 0$:

$$\frac{d\dot{V}}{dh} = -\left(\frac{2\rho_0 B}{\beta \sin\gamma_e}(-\beta) e^{-\beta h} - \beta\right) \rho_0 B V_e^2 \exp\left\{\frac{2\rho_0 B}{\beta \sin\gamma_e} e^{-\beta h} - \beta h\right\} = 0$$

that is

$$\frac{2\rho_0 B}{\sin\gamma_e} e^{-\beta h} + \beta = 0$$

or

$$e^{-\beta h} = -\frac{\beta \sin \gamma_e}{2\rho_0 B} \tag{4.13}$$

The above is a formula for the altitude h^* where the peak of deceleration occurs:

$$h^* = \frac{1}{\beta} \ln \left(-\frac{2\rho_0 B}{\beta \sin \gamma_e} \right) \tag{4.14}$$

The velocity $V^* = V(h^*)$ corresponding to the peak of deceleration is computed using Eq. (4.11) with $h = h^*$ of formula Eq. 4.14, so that:

$$V^* = V(h^*) = V_e \exp \left\{ \frac{\rho_0 B}{\beta \sin \gamma_e} e^{-\beta h^*} \right\} = \frac{V_e}{\sqrt{e}} = 0,607\ V_e \tag{4.15}$$

To compute the peak of deceleration, it is possible to use (4.12), (4.13) and (4.14) to get:

$$a_{max} = \dot{V}(h^*) = \frac{\beta}{2\,e} V_e^2 \sin \gamma_e \tag{4.16}$$

It is important to note that the maximum deceleration does not depend on the geometry and the mass of the capsule, but only on the state conditions at the entry point V_e, γ_e.

Figure 4.12 shows the velocity and the deceleration behaviour with respect to the altitude according to the Allan-Egger formulas. The point marked by "*" indicates the velocity at which the maximum deceleration occurs. The plots show the results for altitude in the range 100 Km to 50 Km, that is the range of applicability of the formulas. The upper limit in altitude is given by the point where the drag begins to exceed the gravity, the lower limit is constrained by the assumption of constant drag coefficient.

Peak of Heat

In the 1950s, initial research into the re-entry problem focused on slender bodies. However, wind tunnel tests showed that a lot of heat was transferred to the surface of the vehicle, which could not withstand the heat flow. In 1952, H. Julian Allan, a research scientist at the Ames Aeronautical Laboratory of the National Advisory Committee for Aeronautics (NACA), made a curious discovery: by increasing the drag of the vehicle, it was possible to reduce the heat transferred to the vehicle. In fact, most of the heat flow was redirected away from the car's surface. The first real designs were then made by Allan with a second researcher Alfred J. Eggers, who proposed "blunt bodies" for re-entry capsules. These were able to generate shock waves that were detached from the surface and placed in front of the body, thus removing heat from the vehicle.

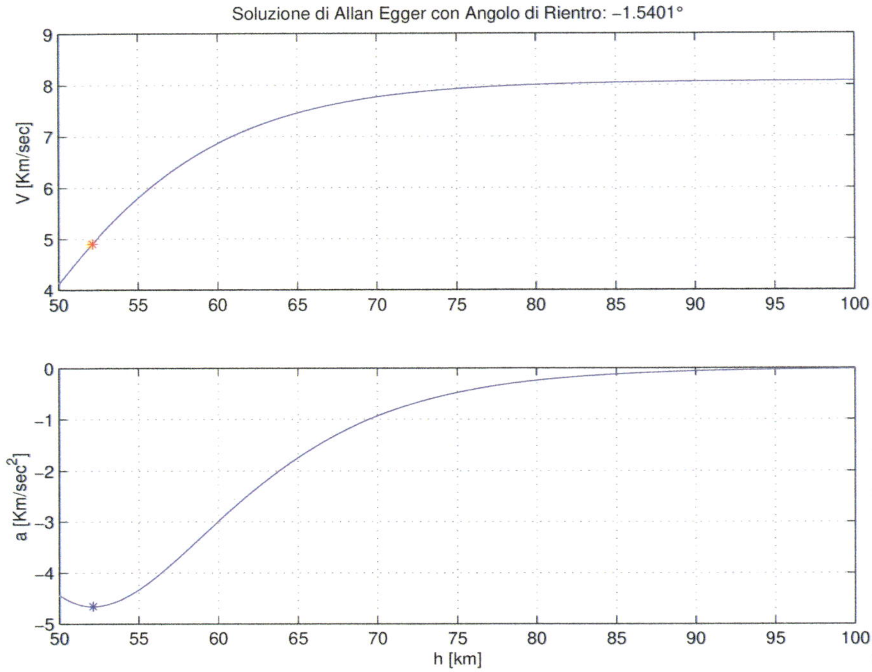

Fig. 4.12 The velocity (in km/s) during a ballistic reentry trajectory as function of the altitude

The following formula was introduced for the incoming heat flux [9]:

$$q = \frac{\mathrm{C}}{\sqrt{\mathrm{R}}} \sqrt{\frac{\rho}{\rho_0}} \frac{V^3}{V_c^3} \quad \left[\frac{W}{m^2}\right] \tag{4.17}$$

with C = 20 281 to have the heat flux in $\frac{KW}{m^2}$, R is the radius of curvature of the heat shields at stagnation point (for instance equal to 0.3 m for Soyuz capsule), ρ_0 atmospheric density at Earth surface and $V_c = \sqrt{\mu/R_\oplus}$.

To design the heat shield it is necessary to compute the maximum temperature that may occur during the reentry. The heat shield can be made of very high temperature resistant or ablative materials. A rather conservative estimate of the maximum temperature is obtained considering the thermal equilibrium between the incoming thermal flux and the outgoing flux radiated by the heat shield:

$$q = \varepsilon\, \sigma\, T^4$$

where $\sigma = 5.67e - 8\frac{W}{m^2\,K^4}$ is the Boltzmann constant, ε is the surface emissivity depending on the material: for instance the Titanium ($Ti - 5\,Al$) has $\epsilon = 0.4$ with working temperature up to $\mathrm{T}_{max} = 1400C = 1673\,K$.

The equilibrium condition gives the temperature at surface:

$$T = \left(\frac{q}{\varepsilon\,\sigma}\right)^{\frac{1}{4}} \leq \mathrm{T}_{max}$$

In particular the following constraint must be taken into account:

$$\left(\frac{q_{max}}{\varepsilon\,\sigma}\right)^{\frac{1}{4}} \leq \mathrm{T}_{max}$$

To compute the maximum thermal flow q_{max}, it is possible to use in (4.17) the double exponential formula of the velocity [9]

$$q = \frac{\mathrm{C}}{\sqrt{\mathrm{R}}}\,\frac{V_e^3}{V_c^3}\exp\left\{3\frac{\rho_0\,B}{\beta\,\sin\gamma_e}\,e^{-\beta h} - \beta\,\frac{h}{2}\right\} \tag{4.18}$$

The peak of heat occurs at the altitude h^{**} such that $\frac{dq}{dh} = 0$ and from (4.18):

$$h^{**} = \frac{1}{\beta}\,\ln\left(-\frac{6\rho_0 B}{\beta\,\sin\gamma_e}\right) \tag{4.19}$$

The velocity at the peak of heat is equal to $V^{**} = V(h^{**})$, that is:

$$V^{**} = V_e\,\exp\left\{\frac{\rho_0 B}{\beta\,\sin\gamma_e}e^{-\beta\,h^{**}}\right\} = V_e\,e^{-\frac{1}{6}} \simeq 0.8465\,V_e \tag{4.20}$$

The value of the maximum heat flow is obtained by the formula of V^{**} in (4.17), and h^{**} in (4.18), and the result is:

$$q_{max} = \frac{\mathrm{C}}{\sqrt{e\mathrm{R}}}\left(\frac{V_e}{V_c}\right)^3\sqrt{-\frac{\beta\,\sin\gamma_e}{6\rho_0\,B}} \tag{4.21}$$

Note that:

- The peak of heat occurs before the peak of deceleration, since it is easy to note that $h^{**} > h^*$.
- As opposite to the peak of deceleration, the peak of heat depends on the capsule ballistic coefficient B and decreases with increasing ballistic coefficient and curvature radius R.

Table 4.2 shows the reentry constraints on a capsule during reentry.

Figure 4.13 shows the numerical output and the peak of deceleration and flux obtained by the Allan Egger formulas and denoted by '*'. The numerical integration starts at an altitude of 300 km and flight path angle –5 deg. The peaks obtained by the analytic formulas are rather close to the numerical ones.

Table 4.2 Limits on deceleration and heat flow

	Formula by [9]	Constraints
a_{max}	$\frac{\beta}{2e} V_e^2 \sin\gamma_e$	$\leq 6\,g$(with crew)
		$\leq 15\,g$ (unmanned)
q_{max}	$\frac{C}{\sqrt{eR}} \left(\frac{V_e}{V_c}\right)^3 \sqrt{-\frac{\beta \sin\gamma_e}{6\rho_0 B}}$	$\leq \varepsilon\,\sigma\,T_{max}^4$ (heat shield)

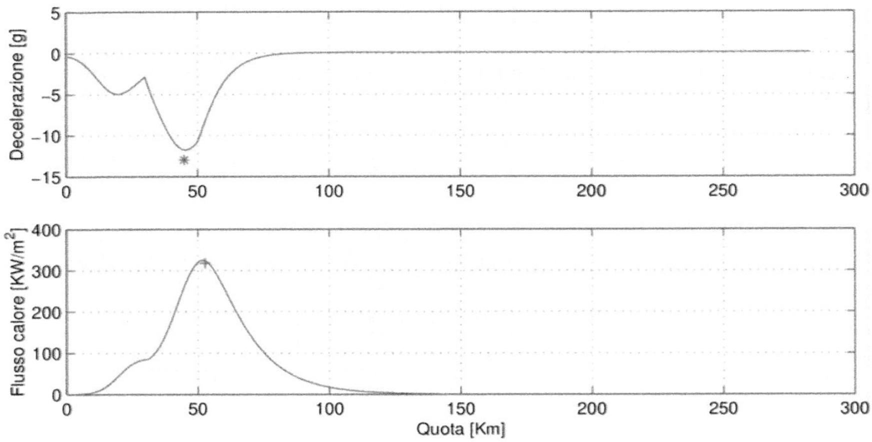

Fig. 4.13 Deceleration (in g) and thermal flux (in KW/m^2) for a Soyuz like reentry capsule from 300 km altitude and flight path angle equal to 5 deg. The symbols '*' show the peak deceleration and peak of heat defined by the analytic formulas

4.2.1 The Composite Solution

The Allen-Eggers solution depends on the reentry conditions (V_e, γ_e, h_e) corresponding to the condition of drag resistance greater than the gravity. However this is a rather vague statement: the reentry condition can be fixed at altitude between 150/100 km, but this depends also on the velocity of the capsule as it approaches the Earth. Moreover it is useful to compute the conditions of the entry corridor with respect to the initial keplerian conditions, also for the design of the reentry manoeuvre.

Consider for instance the reentry of a capsule from the International Space Station (ISS) at 300 km of altitude.

After a sequence of low thrust manoeuvres to safely separate the reentry capsule from the ISS, a high thrust braking manoeuvre is performed at a safe distance and altitude a bit smaller than 300 km.

After such braking ΔV the capsule is on an elliptic trajectory passing through the high density layers of the atmosphere to reentry to Earth (direct reentry). It is important to design such a ΔV to achieve a safe reentry trajectory within the reentry corridor.

In fact to properly design such a ΔV it is necessary to have a solution that holds within the "keplerian region" and not just within the atmospheric region. Outside the atmosphere the solution is the well known keplerian orbit, in the atmospheric region the Allen-Eggers solution applies. In the following the two solutions are matched.

Let the independent variable r be used in the reentry equations (4.1):

$$\begin{aligned}\dot{V} &= \frac{dV}{dt} = \frac{dV}{dr}\frac{dr}{dt} = \frac{dV}{dr}V\sin\gamma = \frac{1}{2}\frac{dV^2}{dr}\sin\gamma \\ \dot{\gamma} &= \frac{d\gamma}{dt} = \frac{d\gamma}{dr}\frac{dr}{dt} = \frac{d\gamma}{dr}V\sin\gamma = -\frac{d\cos\gamma}{dr}V\end{aligned}$$

hence

$$\begin{cases}\dfrac{dV^2}{dr} = -\dfrac{2\mu}{r^2} - \dfrac{2\rho B V^2}{\sin\gamma} \\ \dfrac{d\cos\gamma}{dr} = \left(\dfrac{\mu}{r^2V^2} - \dfrac{1}{r}\right)\cos\gamma\end{cases} \tag{4.22}$$

The parameter $\epsilon = \frac{1}{\beta R_\oplus}$ is now introduced to define the different regions of validity of the two different solutions: the keplerian and the Allan-Eggers one. The density formula can be written in terms of the parameter ϵ:

$$\rho(\epsilon) = \rho_0 e^{-\beta(r-R_\oplus)} = \frac{\rho_0}{e^{\frac{1}{\epsilon}\left(\frac{r}{R_\oplus}-1\right)}}$$

Note that for $\epsilon \longrightarrow 0$ the density goes to zero then the drag disappears and we are in the keplerian region.

The Eq. 4.22 become:

$$\begin{cases}\dfrac{dV^2}{dr} = -\dfrac{2\mu}{r^2} - \dfrac{2\rho(\epsilon) B V^2}{\sin\gamma} \\ \dfrac{d\cos\gamma}{dr} = \left(\dfrac{\mu}{r^2V^2} - \dfrac{1}{r}\right)\cos\gamma\end{cases} \tag{4.23}$$

and for $\epsilon \to 0$ become:

$$\frac{dV^2}{dr} = -\frac{2\mu}{r^2} \tag{4.24}$$

$$\frac{d\cos\gamma}{dr} = \left(\frac{\mu}{r^2V^2} - \frac{1}{r}\right)\cos\gamma \tag{4.25}$$

The above equation can be integrated analytically:

$$\int_{V_0}^{V} dV^2 = -\int_{r_0}^{r} \frac{2\mu}{r^2} dr$$

then

$$V^2 = \left(V_0^2 - \frac{2\mu}{r_0}\right) + \frac{2\mu}{r}$$

corresponding to the conservation of the keplerian energy. Namely introducing the constant value of the keplerian energy $\mathcal{E} = \frac{V^2}{2} - \frac{\mu}{r}$, one has

$$V^2 = 2\mathcal{E}_0 + \frac{2\mu}{r} \tag{4.26}$$

Using this equation in (4.25) the following equation is obtained

$$\frac{d\cos\gamma}{\cos\gamma} = \left(\frac{\mu}{2\mathcal{E}_0 r^2 + 2\mu r} - \frac{1}{r}\right) dr$$

Integrating:

$$\ln\frac{\cos\gamma}{\cos\gamma_0} = \frac{\mu}{2}\left(-\frac{1}{\mu}\ln\frac{\mathcal{E}_0 r + \mu}{\mathcal{E}_0 r_0 + \mu} + \frac{1}{\mu}\ln\frac{r}{r_0}\right) - \ln\frac{r}{r_0}$$

With easy steps one gets:

$$\cos\gamma = \cos\gamma_0 \sqrt{\frac{\mathcal{E}_0 r_0^2 + \mu r_0}{\mathcal{E}_0 r^2 + \mu r}} \tag{4.27}$$

In summary, for $\epsilon \to 0$ the keplerian solution was obtained, with velocity given in (4.26) and flight path angle (4.27).

Now the variable $\tilde{r}$ is defined by:

$$r = R_\oplus(1 + \epsilon ln\epsilon) + \epsilon R_\oplus \tilde{r}$$

Transforming the equation of motion in the $\tilde{r}$ variable it will be shown that the limit $\epsilon \to 0$ generates the Allen-Eggers solution.

Note that for $\lim_{\epsilon\to 0} r(\epsilon) = R_\oplus$,that is we are close to the Earth surface for small ϵ. The system of equations (4.23) is, in the new variable $\tilde{r}$:

$$\frac{dV^2}{dr} = \frac{dV^2}{d\tilde{r}}\frac{d\tilde{r}}{dr} = \frac{dV^2}{d\tilde{r}}(\epsilon R_\oplus)^{-1}$$

and with easy steps one obtains

$$\begin{cases} \dfrac{dV^2}{d\tilde{r}} = \epsilon R_\oplus \dfrac{dV^2}{dr} \\[2ex] \dfrac{d\cos\gamma}{d\tilde{r}} = \epsilon R_\oplus \dfrac{d\cos\gamma}{dr} \end{cases} \tag{4.28}$$

Introducing these formulas in (4.23) one has:

$$\begin{cases} \dfrac{dV^2}{d\tilde{r}} = -\dfrac{2\epsilon R_\oplus \mu}{r(\epsilon)^2} - \dfrac{2\epsilon R_\oplus \rho(\epsilon) B V^2}{\sin\gamma} \\ \dfrac{d\cos\gamma}{d\tilde{r}} = \epsilon R_\oplus \left(\dfrac{\mu}{r(\epsilon)^2 V^2} - \dfrac{1}{r(\epsilon)} \right) \cos\gamma \end{cases} \tag{4.29}$$

Writing the expression $\epsilon\rho(\epsilon)$ in function of di $\tilde{r}$:

$$\epsilon\rho(\epsilon) = \epsilon\rho_0 e^{-\frac{1}{\epsilon}(\frac{r(\epsilon)}{R_\oplus}-1)} = \epsilon\rho_0 e^{-\frac{1}{\epsilon}(1+\epsilon\ln\epsilon+\epsilon\tilde{r}-1)} = \rho_0 \epsilon e^{-\ln\epsilon} e^{-\tilde{r}}$$

Then, in the limit

$$\lim_{\epsilon\to 0} \rho_0 \epsilon e^{-\ln\epsilon} e^{-\tilde{r}} = \rho_0 e^{-\tilde{r}}$$

and with this limit the system (4.29) becomes:

$$\begin{cases} \dfrac{dV^2}{d\tilde{r}} = -2R_\oplus \rho_0 \dfrac{e^{-\tilde{r}} B V^2}{\sin\gamma} \\ \dfrac{d\cos\gamma}{d\tilde{r}} = 0 \end{cases} \tag{4.30}$$

From the second equation it follows $\gamma = \gamma_e$, while the first equation can be integrated analytically:

$$\frac{dV^2}{V^2} = -\frac{2R_\oplus \rho_0 B}{\sin\gamma_e} e^{-\tilde{r}} d\tilde{r}$$

so

$$\ln \frac{V^2}{V_e^2} = \frac{2R_\oplus \rho_0 B}{\sin\gamma_e} \left[e^{-\tilde{r}} \right]_{r_e}^{r} \simeq \frac{2R_\oplus \rho_0 B}{\sin\gamma_e} e^{-\tilde{r}}$$

hence

$$V = V_e \exp\left\{ \frac{R_\oplus \rho_0 B}{\sin\gamma_e} e^{-\tilde{r}} \right\}$$

If written in the variable r the above solution is just the Allan-Eggers solution.

$$V_{AE} = V_e \exp\left\{ \frac{R_\oplus \rho_0 B}{\sin\gamma_e} e^{-\tilde{r}(r)} \right\}$$

Therefore two solutions are available: the Kepler solution with initial conditions (V_0, γ_0) and the Allen-Eggers solution (V_{AE}, γ_{AE}). The two different solutions are derived by the same system of differential equation putting $\epsilon \to 0$, but with different variables, the variable r for Kepler and the variable $\tilde{r}$ for the Allan-Eggers solution.

To get a solution that holds true uniformly with the distance r, the following composed solution is introduced:

$$V_{comp}(r) = V_k(r) + V_{AE}\left(\tilde{r}(r)\right) - \lim_{r\to\infty} V_{AE}\left(\tilde{r}(r)\right) \tag{4.31}$$

In this way at high altitudes the difference

$$V_{AE}(r) - \lim_{r\to\infty} V_{AE}(r)$$

tends to zero and the composite solution (4.31) tends to the keplerian solution V_k.

In the same way, it is required that for $r \to R_\oplus$ the composed solution tends to the solution V_{AE}, as in the formula:

$$V_{comp}(r) = V_{AE}\left(\tilde{r}(r)\right) + V_k(r) - \lim_{r\to R_\oplus} V_k(r) \tag{4.32}$$

Comparing Eqs. 4.31 and 4.32 it follows that the two relationships are both true if and only if the following matching condition is satisfied:

$$\lim_{r\to\infty} V_{AE}\left(\tilde{r}(r)\right) = \lim_{r\to R_\oplus} V_k(r) \tag{4.33}$$

The same arguments hold true in matching the two formulas of the flight path angle:

$$\lim_{r\to\infty} \gamma_{AE}\left(\tilde{r}(r)\right) = \lim_{r\to R_\oplus} \gamma_k(r) \tag{4.34}$$

The consequence of the matching condition (4.33) is found computing the limit of (4.11):

$$\lim_{r\to\infty} V_{AE}\left(\tilde{r}(r)\right) = V_e$$

and the limit of Eq. 4.26:

$$\lim_{r\to R_\oplus} V_k(r) = \sqrt{2\mathcal{E}_0 + \frac{2\mu}{R_\oplus}}$$

Then the matching condition on the velocity is:

$$V_e = \sqrt{2\mathcal{E}_0 + \frac{2\mu}{R_\oplus}} \tag{4.35}$$

At same the matching condition on the flight path angle is computed recalling that the Allan-Eggers solution implies a constant flight path angle and this is also the limit for $r \to \infty$, whereas close to the Earth surface the limit of Eq. 4.27 is:

$$\lim_{r \to R_\oplus} \cos \gamma_k(r) = \cos \gamma_0 \sqrt{\frac{\mathcal{E}_0 r_0^2 + \mu r_0}{\mathcal{E}_0 R_\oplus^2 + \mu R_\oplus}}$$

hence the matching condition on the flight path angle implies:

$$\cos \gamma_e = \cos \gamma_0 \sqrt{\frac{\mathcal{E}_0 r_0^2 + \mu r_0}{\mathcal{E}_0 R_\oplus^2 + \mu R_\oplus}} \tag{4.36}$$

The matching conditions (4.35), (4.36) allow to write the entry conditions (V_e, γ_e) as function of the initial condition (V_0, γ_0), in fact the composed solution can be written as function of the initial condition (V_0, γ_0) only:

$$V_{comp} = \sqrt{2\mathcal{E}_0 + \frac{2\mu}{r}} + \sqrt{2\mathcal{E}_0 + \frac{2\mu}{R_\oplus}} \left(\exp\left\{ \frac{\rho_0 B}{\beta \sin \gamma_e} e^{-\beta h} \right\} - 1 \right) \tag{4.37}$$

where, from Eq. 4.36 it follows:

$$\sin \gamma_e = -\sqrt{1 - \cos^2 \gamma_e}$$

with $\cos \gamma_e$ in (4.36).

The composed solution for the flight path angle is equal to the keplerian value given by (4.27):

$$\cos \gamma_{comp} = \cos \gamma_k + \cos \gamma_e - \cos \gamma_e = \cos \gamma_k \quad \text{that is}$$

$$\cos \gamma_{comp} = \cos \gamma_0 \sqrt{\frac{\mathcal{E}_0 r_0^2 + \mu r_0}{\mathcal{E}_0 r^2 + \mu r}} \tag{4.38}$$

The composed solutions (4.37), (4.38) allow to consider the analytic results of the reentry dynamics in terms of the initial conditions, when the capsule is far away from the atmospheric arch of the trajectory.

For instance the matching condition of Eq. 4.36 implies:

$$\cos \gamma_0 \sqrt{\frac{\mathcal{E}_0 r_0^2 + \mu r_0}{\mathcal{E}_0 R_\oplus^2 + \mu R_\oplus}} \leq 1 \tag{4.39}$$

This relation can be regarded as an equation on V_0 or r_0 and γ_0 to get a direct reentry. The formula of maximum deceleration of (4.16) is now equal to:

$$a_{max} = -\frac{\beta}{2e}\left(2\mathcal{E}_0 + \frac{2\mu}{R_\oplus}\right)\sqrt{1 - \cos^2\gamma_0 \frac{\mathcal{E}_0 r_0^2 + \mu r_0}{\mathcal{E}_0 R_\oplus^2 + \mu R_\oplus}} \tag{4.40}$$

Formula (4.40) determines the maximum deceleration in function of the initial conditions r_0, V_0, γ_0 (and not on the entry conditions r_e, V_e, γ_e which are ambiguously defined).

In the same way the peak of heat given by Eq. 4.21, is now a function of the keplerian initial conditions:

$$q_{max} = \frac{\mathrm{C}}{\sqrt{e}\,\mathrm{R}}\left[\frac{2\mathcal{E}_0 + \frac{2\mu}{R_\oplus}}{\frac{\mu}{R_\oplus}}\right]^{\frac{3}{2}}\sqrt{\frac{\beta}{6\rho_0 B}}\sqrt{1 - \cos^2\gamma_0 \frac{\mathcal{E}_0 r_0^2 + \mu r_0}{\mathcal{E}_0 R_\oplus^2 + \mu R_\oplus}} \tag{4.41}$$

Figure 4.14 compares the velocity profile with the altitude using the numerical and the composite solution for a capsule starting from 300 km of altitude and flight path angle equal to –15 deg. The two trajectories are indistinguishable from each other and the composite solution shows its validity along the whole trajectory. Figures 4.15, 4.16 show the deceleration and the heat flow using the three different methods of computation: the numerical solution, the Allan Egger formulas for the peak of load and peak of heat defined from the numerical entry point at altitude 100 km and the composite solution, defined from the initial Kepler condition. Again the composite solution is close to the numerical integration plot and also approaches the peak values better than the Allan-Egger formulas.

4.2.2 Reentry Corridor

The boundaries of the reentry corridor using the composite solution are now determined.

Considering as initial condition of the reentry capsule the elliptic keplerian orbit obtained after the braking impulse Δu:

$$V_0 = V_C - \Delta u\ ,\ \gamma_0 = 0\ ,\ r_0 = r_a$$

one has that the domain of applicability of the composite formula is (see formula 4.39)

$$Ratio(r_0, V_0) = \sqrt{\frac{\mathcal{E}_0 r_0^2 + \mu r_0}{\mathcal{E}_0 R_\oplus^2 + \mu R_\oplus}} < 1 \tag{4.42}$$

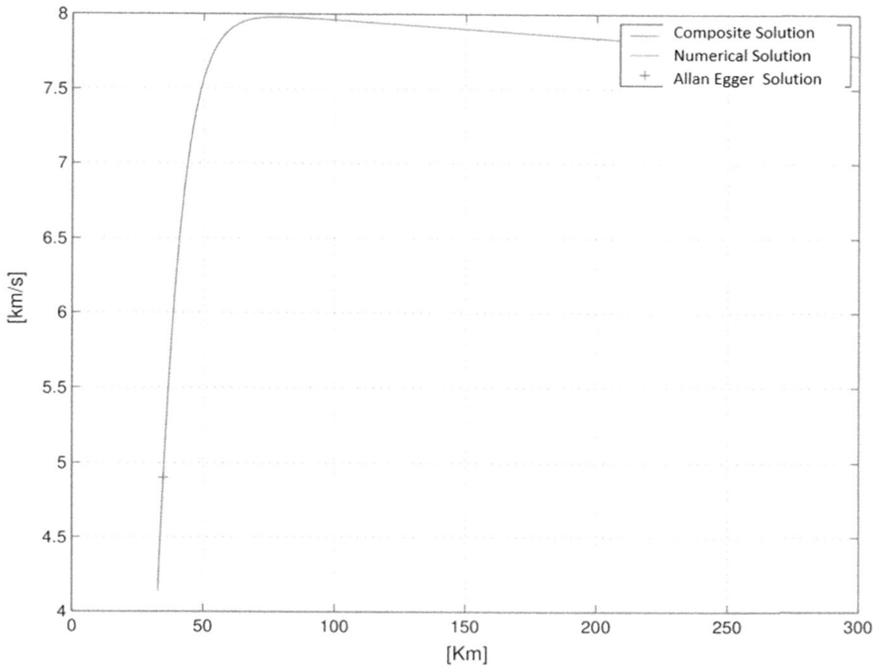

Fig. 4.14 The velocity plot by numerical integration and by the composite solution

This relation only holds if the impulse Δu is large enough. In fact, define the undershot boundary of the reentry corridor as the trajectory starting with radius vector r_0, zero flight path angle and velocity V_0^* such that

$$Ratio(r_0, V_0^*) = 1 \tag{4.43}$$

Then the initial velocity V_0 for the reentry can be selected in the range of velocities $V_0 < V_0^*$, corresponding to different variation of velocity from the velocity V_C of the initial circular orbit, hence to different Δu. Among these velocities there is a limit velocity V_0^{**} which violates deceleration or heat constraints. The trajectory with initial conditions $r_0, \gamma_0 = 0, V_0^{**}$ defines the overshoot boundary of the entry corridor. For instance if $-N\,g$ is the maximum deceleration allowed, the limit velocity is defined by the equation:

$$N\,g = \frac{\beta}{2\,e}\left(2\mathcal{E}_0 + \frac{2\mu}{R_\oplus}\right)\sqrt{1 - \frac{\mathcal{E}_0 r_0^2 + \mu r_0}{\mathcal{E}_0 R_\oplus^2 + \mu R_\oplus}} \tag{4.44}$$

Having fixed r_0, the Eq. 4.44 is an equation in the unique unknown V_0 included in the energy $\mathcal{E}_0$. The solution of Eq. (4.44) is V_0^{**}.

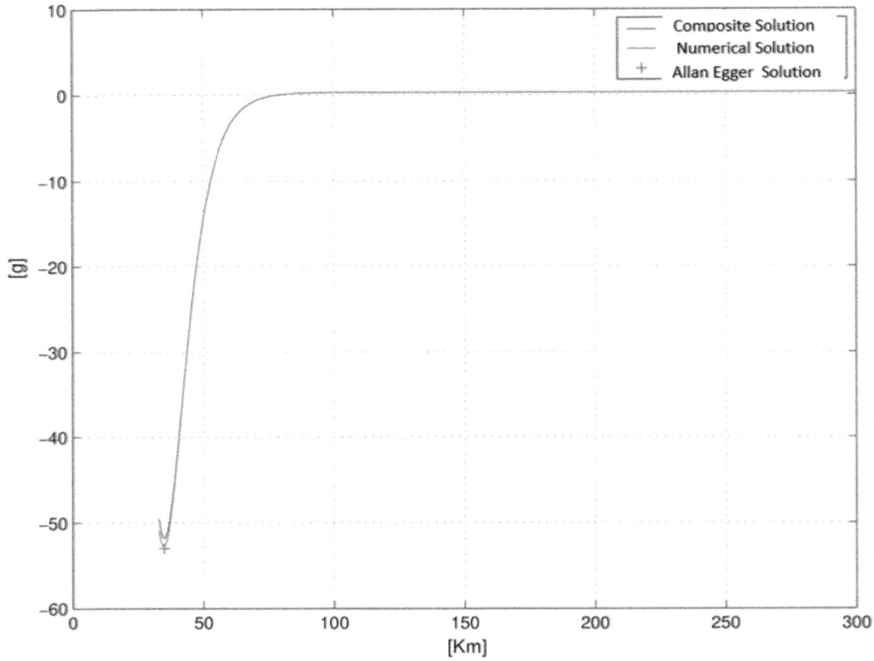

Fig. 4.15 The deceleration plot by numerical integration and by the composite solution. The point denoted by "*" is the peak of load obtained by Allan Egger formula

Thus the entry corridor of a capsule starting at apogee r_0 is the region inside the two trajectories having the initial conditions

$$r_0\ ,\ \gamma_0\ ,\ V_0^*$$

$$r_0\ ,\ \gamma_0\ \ V_0^{**}$$

Table 4.3 shows the values of the entry corridor of the Soyuz ballistic reentry starting from circular orbits of different altitudes: in the second row is reported the value V_0^* corresponding to the undershoot boundary obtained by formula (4.43). Higher velocity will produce a skip trajectory. In the third row it is shown the variation of velocity needed to achieve the velocity V_0^* from the circular orbit $V_C = \sqrt{\frac{\mu}{r_0}}$: $\Delta u_{under} = V_C - V_0^*$. The last two rows reports the velocities V_0^{**} corresponding to the maximum deceleration equal to 15 g and the velocity variation $\Delta u_{over} = V_C - V_0^{**}$.

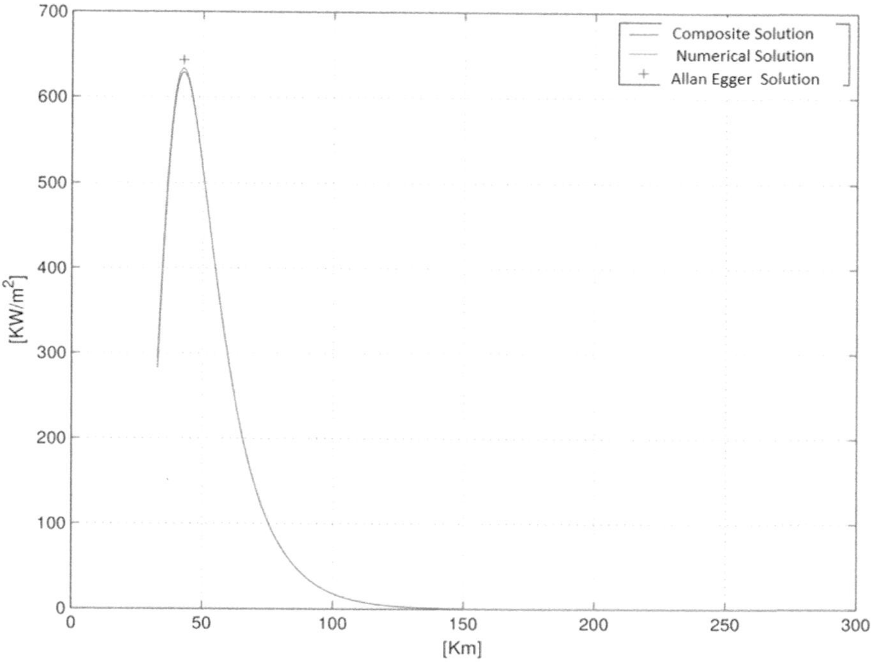

Fig. 4.16 The heat flow plot by numerical integration and by the composite solution. The point denoted by "*" is the peak of heat obtained by Allan Egger formula

Example: Reentry of the Soyuz Capsule from ISS

Consider the Soyuz-like capsule with the same data as the Soyuz but lift coefficient equal to zero. Consider altitude 412 km (such as the ISS altitude) and let us compute its entry corridor according to the above definition.

The velocity of the circular orbit is $V_C = 7.66\,\text{km/s}$. Figure 4.17 shows the function *Ratio* for $r_0 = 412 + R_\oplus$ and for values of the velocity from 7 km/s to V_C. The function *Ratio* is equal to 1 for $V_0 = V_0^* = 7.55\,\text{km/s}$. Figure 4.18 shows the peak of deceleration according to the formula (4.16) for the fixed r_0 and V_0 varying from $6\,\text{km/s}$ to V_0^*. If we fix the maximum deceleration equal to 15 g, one has that the minimum velocity allowed corresponds to $V_0 = V_0^{**} = 7.26\,\text{km/s}$. Therefore, if the capsule Soyuz-like is orbiting with circular orbit of radius $r_0 = 412 + R_\oplus$ then the ΔV to stay within the reentry corridor must be in the interval $\Delta V \in [0.11, 0.40]\,\text{km/s}$.

The Fig. 4.19 show the entry corridor of the capsule starting from the ISS orbit, the undershot boundary is the trajectory with higher range distance obtained with the initial condition: apogee altitude equal to $h_0 = 412\,\text{km}$ and velocity V_0^*. The overshot boundary has the same apogee and initial velocity V_0^{**}.

Table 4.3 The velocities defining the entry corridor related to $a_{max} = 15\,g$ for different apogee altitudes, Soyuz ballistic reentry

h_0 (km)	V_0^* (m/s)	Δu_{under} (m/s)	V_0^{**} (m/s)	Δu_{over} (m/s)
300	7600	126	7250	476
400	7590	79.	7270	399
500	7500	113	7250	363
600	7400	158	7210	348
700	7300	204	7160	344
800	7200	252	7100	352
900	7190	210	7040	360
1000	7100	250	6980	370
1100	7000	301	6910	391
1200	6900	353	6850	402
1300	6890	315	6790	415
1400	6800	359	6720	438
1500	6700	413	6660	453

Note that the entry corridor of Fig. 4.19 corresponds to the constraint of maximum deceleration equal to 15 g, which is not feasible for a crew: for a manned flight the limit of maximum deceleration can be set to 6 g. As a result lower Δu_{over} are allowed, V_0^{**} becomes closer to V_0^* and the entry corridor becomes really narrow. Therefore

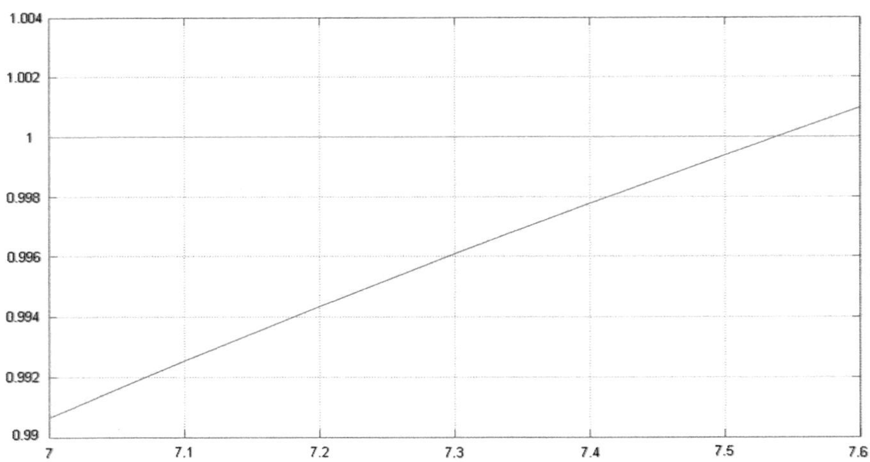

Fig. 4.17 Function $Ratio$ for the ISS orbit. Reentry velocities verify $Ratio < 1$

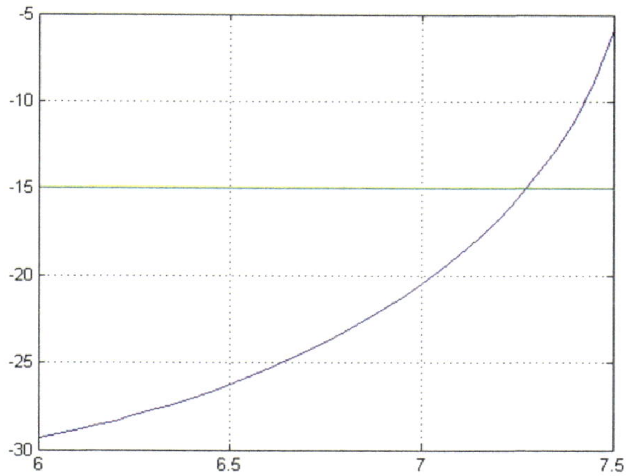

Fig. 4.18 Peak of deceleration for Soyuz starting from ISS orbit for different velocities

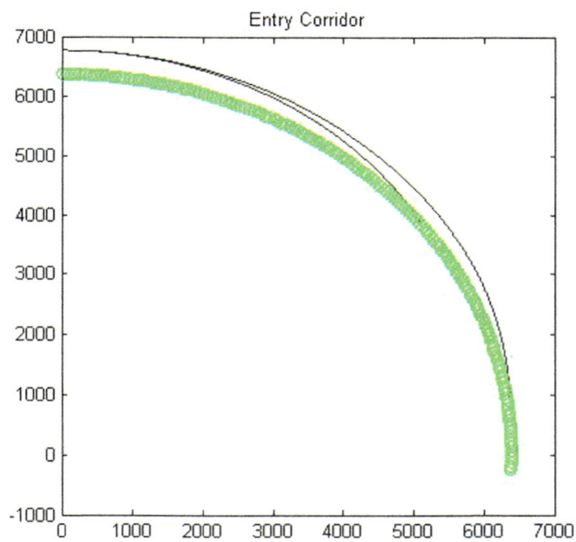

Fig. 4.19 Entry corridor for Soyuz capsule starting from ISS

a small error in the thrust can generate either a skip trajectory or a violation of the peak of deceleration. To overcome this problem it is necessary to add an extra degree of freedom provided by the lift. The lift, acting on the flight path angle variation allows to reduce both the deceleration and the heat flow and can be used as a control variable.

4.3 Reentry with Lift

The (planar) equation of motion taking into account the lift force are

$$\begin{cases} \dot{V} = -\dfrac{\mu}{r^2}\sin\gamma - \dfrac{D}{m} \\ \dot{\gamma} = -\dfrac{\mu}{V\,r^2}\cos\gamma + \dfrac{V}{r}\cos\gamma + \dfrac{L}{m} \\ \dot{r} = V\sin\gamma \\ \dot{\theta} = \dfrac{V}{r}\cos\gamma \end{cases} \tag{4.45}$$

To emphasize the effect of lift, the ballistic reentry ($C_L = 0$) of the Soyuz capsule is compared with the trajectory that would be followed using the correct lift coefficient of the capsule ($C_L = 0.349$).

The standard mission of the Soyuz capsule from the ISS orbit has entry conditions at 110 km of altitude equal to:

$$V_e = 7.9\ \mathrm{km/s},\ \gamma_e = -1.5\,\mathrm{deg},\ h_e = 110\,\mathrm{km}$$

In the case of no lift the profile of the velocity, flight path angle, altitude, deceleration and heat flux are shown in Figs. 4.20, 4.21, 4.22 and 4.23. Taking into account the lift of the capsule, the same variables are in Figs. 4.23, 4.24, 4.25, 4.26 and 4.27.

The following considerations are in order:

- The velocity change is rather smooth and slow with respect to the ballistic case (compare Figs. 4.20 and 4.24)
- The flight path angle is constant for most part of the flight. Then the flight is almost horizontal (the low initial value of the flight is kept constant for long time) (compare Figs. 4.21 and 4.25).
- The peak of declaration is reduced to the value compatible with the presence of a crew (compare Figs. 4.22 and 4.26).
- The heat flux is almost halfed (compare Figs. 4.23 and 4.27).

The lift keeps the reentry capsule on a rectilinear flight with very low slope for long time (equilibrium glide), then the deceleration is distributed in time and as a consequence the heat flux is reduced. The equilibrium glide formulae can be used to approximate a large flight and give an indication of the speed variation and heat flux.

The almost equilibrium flight glide is derived with the following approximations [10, 12]:

$$\gamma = 0\ ,\ \dot{\gamma} = 0\ ,\ \frac{\mu}{r^2} = g$$

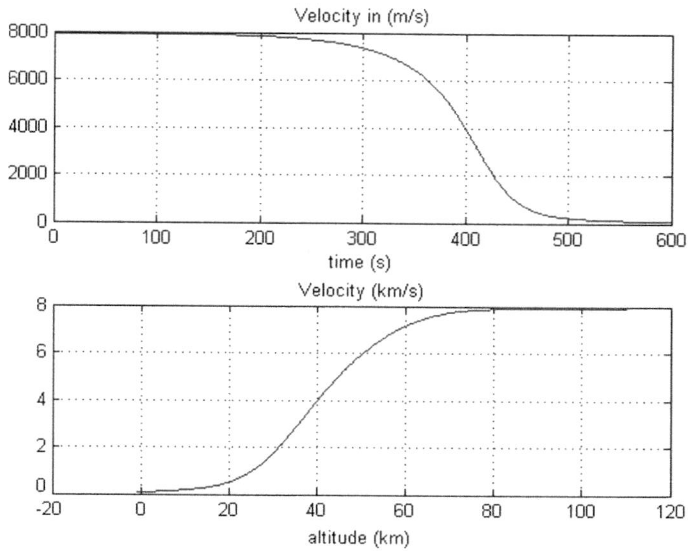

Fig. 4.20 Velocity profile (from reentry altitude) of Soyuz-like capsule (no lift) standard mission from ISS orbit

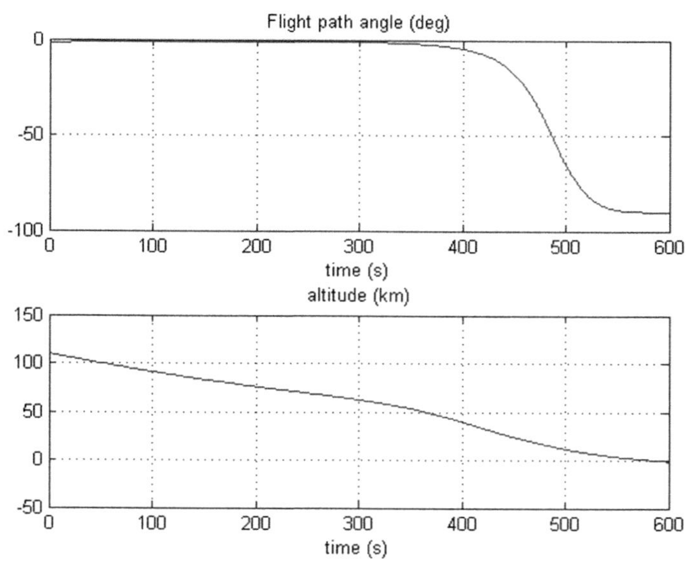

Fig. 4.21 Flight path angle and altitude profile (from reentry altitude) of Soyuz-like capsule (no lift) standard mission from ISS orbit

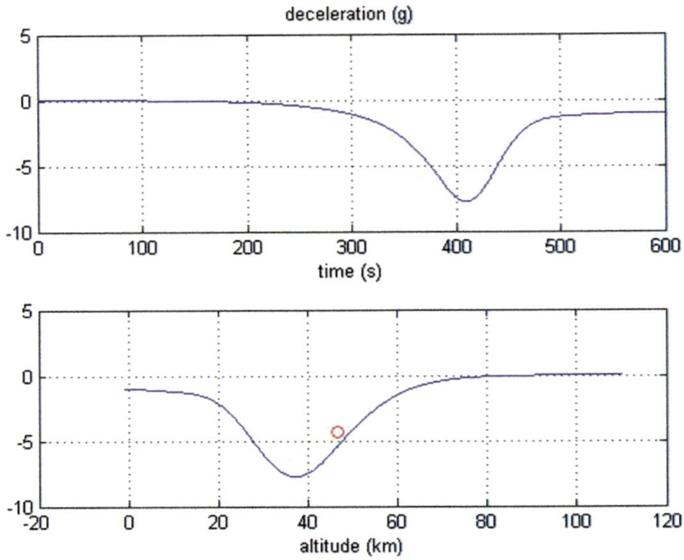

Fig. 4.22 Deceleration profile (from reentry altitude) of Soyuz-like capsule (no lift) standard mission from ISS orbit

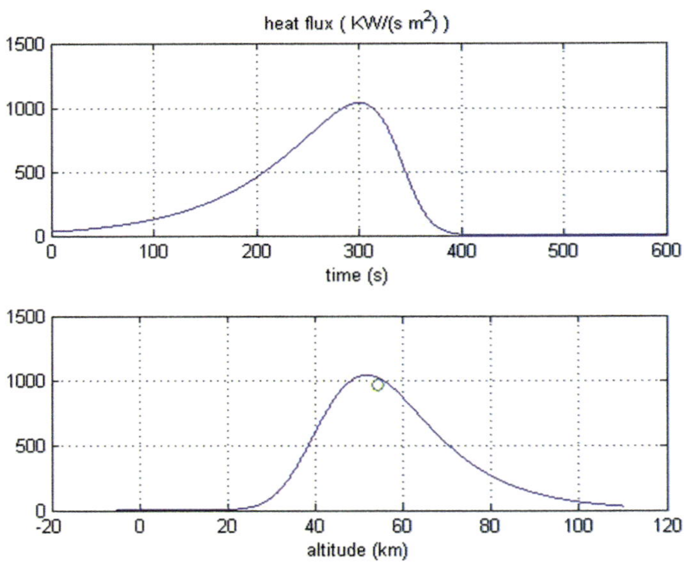

Fig. 4.23 Heat flux profile (from reentry altitude) of Soyuz-like capsule (no lift) standard mission from ISS orbit

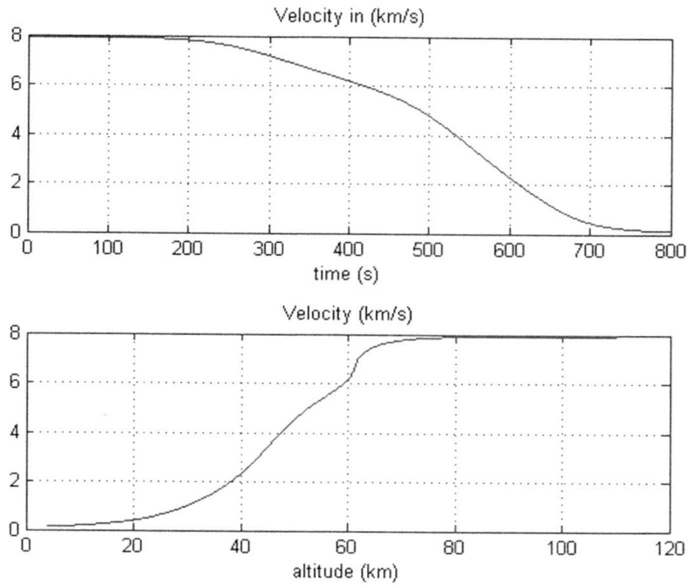

Fig. 4.24 Velocity profile (from reentry altitude) of Soyuz capsule standard mission from ISS orbit

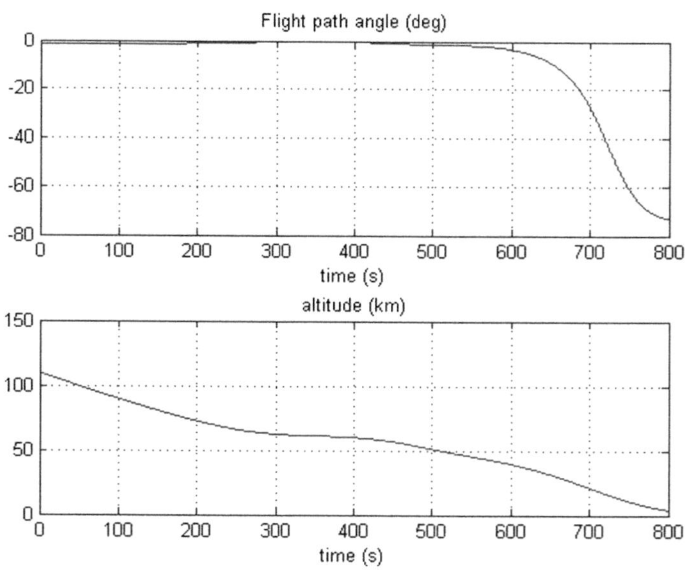

Fig. 4.25 Flight path angle and altitude profile (from reentry altitude) of Soyuz capsule standard mission from ISS orbit

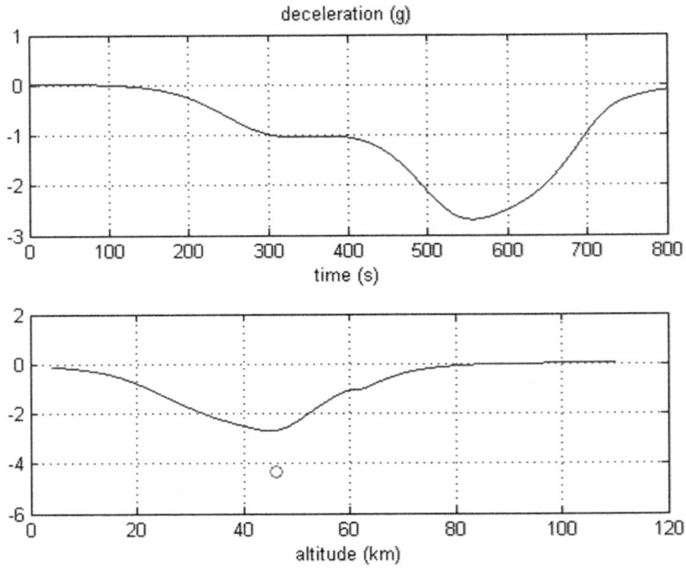

Fig. 4.26 Deceleration profile (from reentry altitude) of Soyuz capsule standard mission from ISS orbit

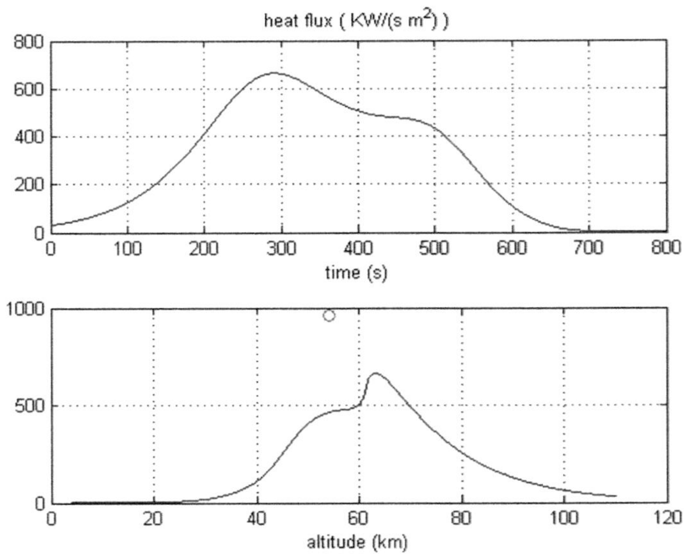

Fig. 4.27 Heat flux profile (from reentry altitude) of Soyuz capsule standard mission from ISS orbit

The variation of velocity is then

$$\dot{V} = -\frac{D}{m}$$

and the flight path angle equation gives:

$$0 = -\frac{g}{V} + \frac{V}{r} + \frac{L}{mV} \tag{4.46}$$

that is

$$\frac{1}{m} = \frac{V}{L}(\frac{g}{V} - \frac{V}{r})$$

Inserting this formula of $\frac{1}{m}$ on the deceleration equation:

$$\dot{V} = -g\frac{D}{L}(1 - \frac{V^2}{g\,r})$$

The further approximation is used: $r = (R_\oplus + h) \sim R_\oplus$ so that

$$\dot{V} = -g\frac{D}{L}(1 - \frac{V^2}{g\,R_\oplus})$$

Note that $g\,R_\oplus = V_E^2$, that is the circular velocity at Earth surface, then:

$$\dot{V} = -\frac{g}{E}(1 - \frac{V^2}{V_E^2})$$

Defining

$$x = \frac{V}{V_E}\ ,\ k = \frac{g}{E}$$

the deceleration equation can be solved by separation of variables:

$$\frac{d\,x}{1 - x^2} = -\frac{k}{V_E}\,dt$$

Integration gives:

$$\int \frac{dx}{1 - x^2} = -\frac{k}{V_E}t + C$$

Since $\frac{1}{1-x^2} = \frac{1}{2}\frac{1}{1-x} + \frac{1}{2}\frac{1}{1+x}$ the integral gives:

$$\int \frac{dx}{1 - x^2} = -\frac{1}{2}log(1 - x) + \frac{1}{2}log(1 + x) = log(\sqrt{\frac{1 + x}{1 - x}})$$

Then

$$log(\sqrt{\frac{1+x}{1-x}}) = -\frac{g}{V_E\, E}\, t + C$$

and

$$\frac{1+x}{1-x} = exp(-\frac{2\,g}{V_E\, E} t + C)$$

Therefore

$$x = \frac{exp(-\frac{2\,g}{V_E\, E} t + C) - 1}{exp(-\frac{2\,g}{V_E\, E} t + C) + 1}$$

and the following formula gives the variation of the velocity during the almost horizontal flight:

$$V = V_E\{\frac{exp(-\frac{2\,g}{V_E\, E} t + C) - 1}{exp(-\frac{2\,g}{V_E\, E} t + C) + 1}\}$$

The constant C is related to the initial condition:

$$\frac{V}{V_E} = \frac{e^C - 1}{e^C + 1}$$

then

$$C = log(\frac{1 + \frac{V_0}{V_E}}{1 + \frac{V_0}{V_E}})$$

hence

$$V = V_E \left\{ \frac{exp(-\frac{2\,g}{V_E\, E} t)(\frac{1+\frac{V_0}{V_E}}{1-\frac{V_0}{V_E}}) - 1}{exp(-\frac{2\,g}{V_E\, E} t)(\frac{1+\frac{V_0}{V_E}}{1-\frac{V_0}{V_E}}) + 1} \right\} \tag{4.47}$$

Note that the variation of the velocity is slower for higher efficiency E, hence for higher lift. The formula (4.47) is now used in the analytic determination of the peak of heat in presence of lift. The heat flux is

$$q = \frac{C}{\sqrt{R}} \sqrt{\frac{\rho}{\rho_E}} (\frac{V}{V_E})^3 \tag{4.48}$$

From the equation of the flight path angle (4.46) it follows that there is equilibrium between the lift, the gravity and centrifugal force:

$$\frac{1}{2}\rho\, C_L S\, V^2 = m\, g\, (1 - \frac{V^2}{g\, r}) \sim m\, g\, (1 - \frac{V^2}{g\, R_\oplus}) = m\, g\, (1 - \frac{V^2}{V_E^2})$$

Then

$$\frac{\rho}{\rho_E} = \frac{m\,g}{\frac{1}{2}\rho_E\,C_L\,S}\,(1-x^2)\,\frac{1}{x^2}\,\frac{1}{V_E} \tag{4.49}$$

and it follows that along the almost horizontal glide the heat flux has equation

$$q = \frac{C}{\sqrt{R}}\sqrt{\frac{m\,g}{\frac{1}{2}\rho_E\,C_L\,S}}\,\sqrt{1-x^2}\,\frac{x^2}{V_E}$$

The peak of the heat flow is computed by the equation

$$\frac{dq}{dx} = 0$$

This equation is equivalent to

$$\frac{d}{dx}\,\sqrt{1-x^2}\,x^2 = 0$$

The solution of the above equation is $x = \sqrt{\frac{2}{3}}$, that is the peak of heat in the case of lift occurs when the velocity has the value

$$V^* = \sqrt{\frac{2}{3}}V_E \sim 0.816V_E$$

(equal to $\sim$ 6.451 km/s in the Soyuz mission).

The peak of heat is therefore obtained inserting the value $x = \sqrt{\frac{2}{3}}$ in Eq. 4.48:

$$q_{max} = \frac{2}{3}\frac{C}{\sqrt{3R}}\sqrt{\frac{m\,g}{\frac{1}{2}\rho_E\,C_L\,S}\,\frac{1}{V_E}} \tag{4.50}$$

Note that the peak of heat depends on the lift as $\sqrt{\frac{1}{C_L}}$, hence the higher lift produces lower heat flux.

Figure 4.28 shows the heat flux of the Soyuz capsule integrated numerically and as function of the velocity. The point 'o' shows the analytical value of the formula (4.28) that is the analytic peak of heat that occurs at velocity $V^* = 6.451$ km/s. There is an error due to the almost horizontal hypothesis, however the result is satisfactory as a first estimate.

In any case, the formulas derived from the equilibrium glide hypothesis clarify the role of the lift in the dynamics of the reentry.

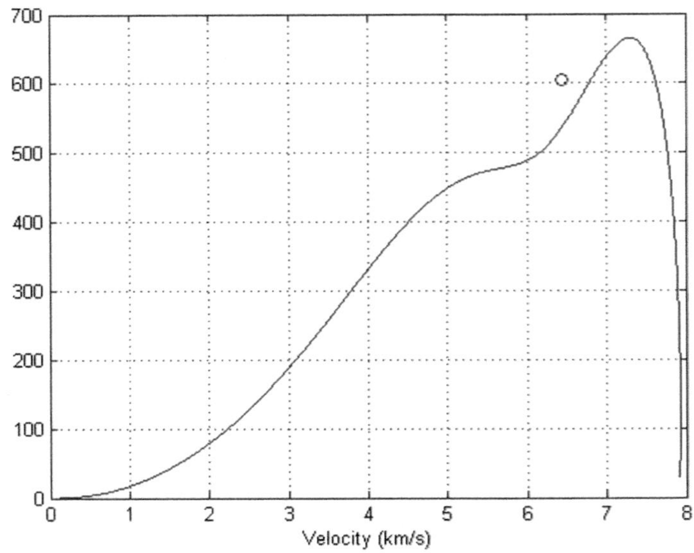

Fig. 4.28 Heat flux profile (from reentry altitude) of Soyuz capsule standard mission from ISS orbit as function of the velocity. The value marked by 'o' is the analytic value of the peak of heat

4.4 Guidance of Reentry Capsules

It was noted that the entry corridor is the fundamental concept in a reentry mission, but another very important point is the targeting of a specific landing area.

In the ballistic case the capsule is basically uncontrolled and its trajectory, and therefore the landing area, depends only on the time and intensity of the manoeuvre performed on the initial (circular) orbit. After the reentry manoeuvre the trajectory can not be corrected: the range and time of flight are determined, and so is the point of impact with the Earth. On the other hand the lift adds an extra dimension to the mission and can be used as control variable to change the range of the reentry capsule. One possible control variable is the bank angle of the capsule. This angle measures the rotation of the capsule with respect to the (relative) velocity and such rotation changes the direction of the flow, as in Fig. 4.29. The figure shows the lift-up and lift-down directions obtained by rotating the capsule. Of course lift-up increases the range of the capsule and the flight time, the lift-down decreases both the range and flight time. Directing the lift in one side of the capsule will steer the capsule out of plane. Banking the capsule also allows to change the lift value in the vertical direction. Namely if ϕ is the bank angle the lift has two components:

Vertical component: $L \cos\phi$

Lateral component: $L \sin\phi$.

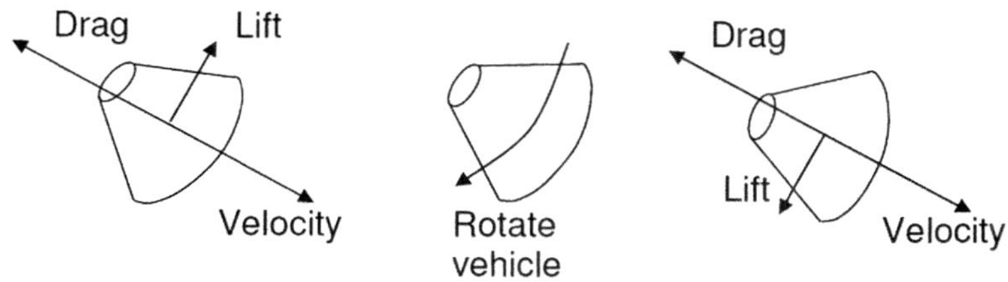

Fig. 4.29 Lift up and lift-down by bank rotation

Fig. 4.30 Lift modulation by periodic bank oscillations

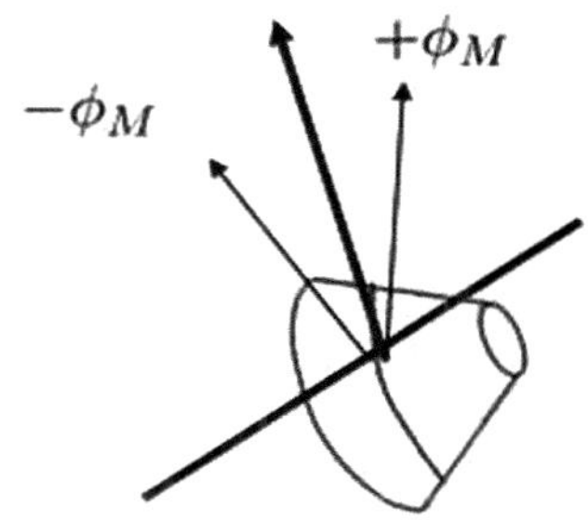

Considering bank oscillations limited between the angles $-\phi_M, +\phi_M$, one has on average (Fig. 4.30):

Average vertical component: $\int_{-\phi_M}^{+\phi_M} L cos\phi \, d\phi = 2L \sin(\phi_M)$

Average lateral component $\int_{-\phi_M}^{+\phi_M} L sin\phi \, d\phi = L \, (cos(\phi_M) - \cos(-\phi_M)) = 0$

Lift modulation can then be achieved by periodic oscillations in bank, realized by a reaction control system producing roll rotations around the longitudinal axis of the capsule by small thrusters. For instance to half the lift modulus along the vertical direction, the amplitude of the bank angle oscillation must be equal to 14.5 deg. On the other hand continuous bank rotations produce zero average lift in all directions, then the capsule's trajectory is essentially ballistic.

Starting from the same entry condition

$$V_e = 7.9 \text{ km/s} \ , \ \gamma_e = -1.5 \text{ deg} \quad h_e = 110 \text{ km} \tag{4.51}$$

different range distances and different time of flight (flight from h_e to 2 km of altitude) are reported according to different bank control law strategies:

- (a) Lift down, that is bank angle $\phi = 180°$.
  Range equal to 2214 km and flight time $t_{f_{ref}} = 400$ s
- (b) Ballistic case, that is bank angle ϕ is rotating continuously.
  Range equal to to 2590 km and flight time $t_{f_{ref}} = 520$ s
- (c) Lift halved , that is bank angle in oscillation with amplitude $\phi = 14.5°$.
  Range equal to 3046 km and flight time $t_{f_{ref}} = 650$ s
- (d) Lift up, that is bank angle $\phi = 0°$.
  Range equal to 3790 km and $t_{f_{ref}} = 810$ s

The range is computed by Eq. 4.45 as $Range = X = R_{\oplus}\, \theta$ or adding to the Eq. 4.45 the range equation:

$$\dot{X} = V\, cos\gamma$$

It is apparent from the above results that bank modulation has a relevant impact on the range.

During the flight different perturbations may act on the capsule changing the nominal trajectory: main source of errors are in the air density models, the definition of aerodynamic characteristics, incidence angle, mass and mass distribution, etc. Then the main data of the nominal trajectory are stored in the on board computer and a correction algorithm is run.

The reentry guidance is split in two phases:

1. Target the nominal reentry state during the initial phase of the aerodynamic control (starting from 0.2 g condition)
2. Track the range distance to the target impact point.

Phase 1

During the phase 1 of the reentry guidance the velocity and flight path angle can be corrected by the variation of C_L to reach as close as possible the nominal entry conditions (for instance (4.51)). The guidance is based upon the formulas of the equilibrium glide that approximates the first arch of the reentry trajectory. At each control time $t = t_0$, with stepsize Δt_1, the current drag D and altitude rate $\dot{h}$ are derived. By these two data the current velocity $V(t_0)$ and the current flight path angle $\gamma(t_0)$ are determined by the formulas:

$$h(t_0) = h(t_{-1}) + \dot{h}\, \Delta t_1 \ , \quad V(t_0) = \sqrt{\frac{D}{\rho\, B}} \ , \quad \gamma(t_0) = asin(\frac{\dot{h}}{V})$$

Then the predicted velocity and altitude are derived by the analytic formulas (4.47), (4.49), till the altitude has reached the entry value, for instance $h_e = 110$ km. If the velocity predicted at the entry altitude is lower than V_e a Lift-down maneuvre is performed, on the other hand if the velocity computed at entry altitude is higher than V_e a Lift-up maneuvre is performed. The algorithm is defined as it follows (ϵ is a tolerance on the entry velocity):

- At time t_0 the current velocity V_0 is computed
- By formula (4.47) compute the estimated velocity V_f at the entry condition h_e
- If $V_f - V_e > \epsilon$ then Lift-up ($\phi = 0°$),
  if $V_f - V_e < -\epsilon$ then Lift-down ($\phi = 180°$)
- $t_0 = t_0 + \Delta t_1$ and iterate.

Phase 2

During the phase 2 of the reentry guidance a reference follower control is implemented mainly to target the impact point which is at latitude Lat and moving toward East at velocity equal to $\omega_E\ R_\oplus\ \cos(Lat)$. The control is on the range and it is updated at fixed time step, for instance $\Delta t_2 = 2$ s. Each two seconds the drag D and rate of descent $\dot{h}$ are derived and the predicted velocity and flight path angle $V(t_0 + \Delta t_2)$, $\gamma(t_0 + \Delta t_2)$ are determined as in Step 1. If at time $t_0 + \Delta t_2$ the predicted state is different from the reference state a different final range $X \neq X_{ref}$ will be reached.

This final range is estimated by the linear perturbations formula:

$$X = X_{ref} + \frac{\partial X}{\partial V}\left(V_{real} - V_{ref}\right) + \frac{\partial X}{\partial \gamma}\left(\gamma_{real} - \gamma_{ref}\right)$$

Then

- If $X - X_{ref}\ > \epsilon$ Lift-down ($\phi = 180°$)
- If $X - X_{ref}\ < \epsilon$ Lift-up ($\phi = 0°$)

where ϵ is a prescribed tolerance.

In the linear perturbations formula the meaning of the partial derivatives is for example: $\frac{\partial X}{\partial V}$ is the change in the range for a difference in the initial velocity. To find this term, let us observe that if x is the solution of the Cauchy problem $\dot{x} = f(x)\ ,\ x(0) = x_0$ then the derivative $\frac{\partial x}{\partial x_0}$ is the solution of the linear system $\dot{\xi} = \frac{\partial f}{\partial x}\,\xi$ with initial condition $\xi(0) = 1$.

To know the variation of range we have to compute the following variations:

$$\frac{\partial X}{\partial V} = R_\oplus \frac{\partial \theta}{\partial V}\ ,\ \frac{\partial X}{\partial \gamma} = R_\oplus \frac{\partial \theta}{\partial \gamma}$$

where

- $\frac{\partial \theta}{\partial V}$ is the fourth component of the solution of the equation obtained by linearizing equation 4.45 integrated till time $t_{f_{ref}}$ with initial condition $(1, 0, 0, 0)$
- $\frac{\partial \theta}{\partial \gamma}$ is the fourth component of the solution of the equation obtained by linearizing equation 4.45 integrated till time $t_{f_{ref}}$ with initial condition $(0, 1, 0, 0)$

The linearized system is:

$$\dot{\mathbf{y}} = \begin{bmatrix} \frac{\partial f_1}{\partial V} & \frac{\partial f_1}{\partial \gamma} & \frac{\partial f_1}{\partial r} & \frac{\partial f_1}{\partial \theta} \\ \frac{\partial f_2}{\partial V} & \frac{\partial f_2}{\partial \gamma} & \frac{\partial f_2}{\partial r} & \frac{\partial f_2}{\partial \theta} \\ \frac{\partial f_3}{\partial V} & \frac{\partial f_3}{\partial \gamma} & \frac{\partial f_3}{\partial r} & \frac{\partial f_3}{\partial \theta} \\ \frac{\partial f_4}{\partial V} & \frac{\partial f_4}{\partial \gamma} & \frac{\partial f_4}{\partial r} & \frac{\partial f_4}{\partial \partial \theta} \end{bmatrix} \mathbf{y} \tag{4.52}$$

with

$$\begin{aligned} f_1 &= -\frac{\mu}{r^2}\sin\gamma - \frac{D}{m} \\ f_2 &= -\frac{\mu}{V r^2}\cos\gamma + \frac{V}{r}\cos\gamma + \frac{L}{m} \\ f_3 &= V\sin\gamma \\ f_4 &= \frac{V}{r}\cos\gamma \end{aligned} \tag{4.53}$$

Then

$$\begin{aligned} \frac{\partial f_1}{\partial V} &= -2\rho\, B\, V \\ \frac{\partial f_1}{\partial \gamma} &= -\frac{\mu}{r^2}\cos\gamma \\ \frac{\partial f_1}{\partial r} &= 2\frac{\mu}{r^3}\sin\gamma + \beta\,\rho B\, V^2 \\ \frac{\partial f_1}{\partial \theta} &= 0 \end{aligned} \tag{4.54}$$

$$\begin{aligned} \frac{\partial f_2}{\partial V} &= \frac{\mu}{r^2 V^2}\cos\gamma + \frac{1}{r}\cos\gamma + \rho\, B\, E\, V \\ \frac{\partial f_2}{\partial \gamma} &= \frac{\mu}{V r^2}\sin\gamma - \frac{V}{r}\sin\gamma \\ \frac{\partial f_2}{\partial r} &= 2\frac{\mu}{r V r^3}\cos\gamma - \frac{V}{r^2}\cos\gamma - \beta\,\rho B\, E\, V^2 \\ \frac{\partial f_2}{\partial \theta} &= 0 \end{aligned} \tag{4.55}$$

$$\begin{aligned} \frac{\partial f_3}{\partial V} &= \sin\gamma \\ \frac{\partial f_3}{\partial \gamma} &= V\cos\gamma \\ \frac{\partial f_3}{\partial r} &= 0 \\ \frac{\partial f_3}{\partial \theta} &= 0 \end{aligned} \tag{4.56}$$

$$\begin{aligned} \frac{\partial f_4}{\partial V} &= \frac{1}{r}\cos\gamma \\ \frac{\partial f_4}{\partial \gamma} &= -\frac{V}{r}\sin\gamma \\ \frac{\partial f_4}{\partial r} &= 0 \\ \frac{\partial f_4}{\partial \theta} &= 0 \end{aligned} \tag{4.57}$$

The above formulas evaluated on the reference state variables V_{ref}, γ_{ref}, h_{ref}, θ_{ref} are the entries of the matrix of equation 4.52. Then the following sequence defines the guidance of the phase 2:

- $t_0 \longrightarrow$ evaluate the current velocity and flight path angle V_0, γ_0
- Compute $\frac{\partial X}{\partial V} = R_\oplus \frac{\partial \theta}{\partial V}$, $\frac{\partial X}{\partial \gamma} = R_\oplus \frac{\partial \theta}{\partial \gamma}$ where $\frac{\partial \theta}{\partial V}$, $\frac{\partial \theta}{\partial \gamma}$ are the solutions y_4 of equation 4.52 integrated from t_0 to $t_{f\,ref} - t_0$ with initial conditions $(1, 0, 0, 0)$, $(0, 1, 0, 0)$ respectively.
- Compute the estimated range: $X = X_{ref}(t_0) + \frac{\partial X}{\partial V}(V_0 - V_{ref}) + \frac{\partial X}{\partial \gamma}(\gamma_0 - \gamma_{ref})$
- Set Lift up if $(X - X_{ref}) < \epsilon$, set Lift down if $(X - X_{ref}) > \epsilon$
- $t_0 = t_0 + \Delta t_2 s$ and iterate.

A similar algorithm can be applied for lateral control, that is if a sensible out of plane deviation is detected, a bank angle reversal ($\pm 90°$) is commanded. The allowed deviation from the desired plane of the trajectory decreases as the capsule is approaching the target impact point. To compute the deviation and apply the lateral control strategy the three dimensional equation of motion must be considered.

4.5 Three Dimensional Entry Trajectories

The three dimensional kinematic equations are written in the $(\hat{r}, \hat{E}, \hat{N})$ frame defined by the absolute longitude λ_a and the latitude L of the radius vector $\mathbf{r}$ from the Earth centre to the capsule centre of mass, see Fig. 4.31. These are derived as in Section 3.2 considering absolute velocity (hence $\omega_E = 0$) and no thrust.

Then we have the 3D equations of the reentry [10]:

$$\begin{cases} \dot{V} = -\frac{\mu}{r^2} \sin\gamma - \rho\, B\, V^2 \\ \dot{\gamma} = \dfrac{V\ \cos\gamma}{r} - \dfrac{\mu}{V\, r^2} \cos\gamma + \rho\, B\, E\, V\ \cos\mu \\ \dot{\xi} = -\dfrac{V\ \cos\gamma\ \cos\xi}{r} \tan L + \rho\, B\, E\, V\, \dfrac{\sin\mu}{\cos\gamma} \\ \dot{r} = V\ \sin\gamma \\ \dot{\lambda} = \dfrac{V\ \cos\gamma\ \cos\xi}{r\ \cos L} \\ \dot{L} = \dfrac{V\ \cos\gamma\ \sin\xi}{r} \end{cases} \tag{4.58}$$

B is the ballistic factor and E is the aerodynamic efficiency. In the reentry regime, they can be assumed constant.

The angle μ is controlled by a suitable guidance law. For $\mu = 0$ the aerodynamic control is only in the plane of the trajectory and there is a small out of plane drift. The reentry trajectory can be guided by the change of the angle of attack α to be considered in the aerodynamic coefficient C_L included in the coefficients B and E. Higher angle of attack increases the lift and can extend the range of the reentry trajectory. As opposite, if the predicted impact point has a range higher than the target impact point, the angle of attack is reduced. The angle of attack is changed rotating the capsule by a torque generated by attitude engines or by a system of moving some internal masses to change the center of mass position with respect to the centre of pressure, thus generating aerodynamic torque.

The same kind of torques can be implemented to generate a rotation of angle μ to control the plane of the reentry trajectory (lateral guidance).

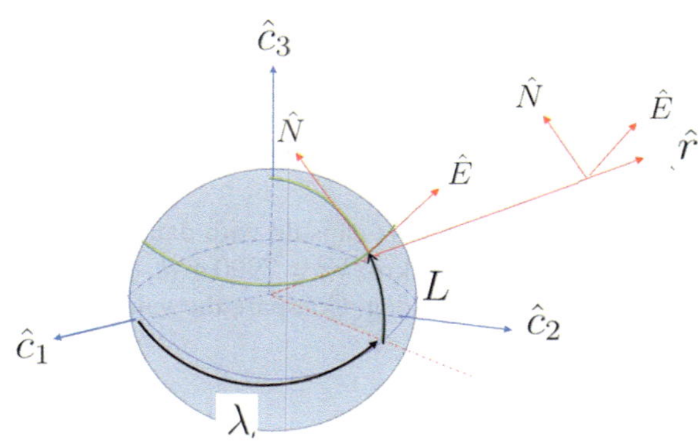

Fig. 4.31 The ECI frame and the moving frame $(\hat{r}, \hat{E}, \hat{N})$

4.6 Lateral Guidance

The target landing point has specific geographic latitude and longitude (L_t, λ_t) at altitude of 10 km[2] and suppose that the current reentry trajectory determines a different landing point (L_f, λ_f). Then the angle μ will be used as control variable to correct the trajectory. As shown in the previous equation the variation of μ will effect both the flight path angle (thus changing the range) and the azimuth angle (thus changing the plane of the trajectory). A possible guidance law is given by

$$\mu = A\, sin(\omega_g\, t) \tag{4.59}$$

where the amplitude A and the frequency ω_g of the bank oscillations are chosen in order to achieve the target landing point.

A two parameter search is implemented to minimize the error

$$Err = \sqrt{(L_t - L_f)^2 - (\lambda_t - \lambda_f)^2}$$

Suppose for instance that the Soyuz capsule has state variables:

$$V_0 = 7.9\,\text{km/s}\,,\ h_0 = 300\,\text{km}\,,\ \gamma_0 = -10.5\,\text{deg}$$

$$\lambda_0 = 45\,\text{deg}\,,\ L_0 = 45\,\text{deg}\,,\ \xi_0 = 28\,\text{deg}$$

and let the target point have geographic coordinates $L_t = 39.8$ deg, $\lambda_t = 62$ deg.

To achieve the target point parameters found are: $A = 2.65$ rad, $\omega = 0.55$ rad/s corresponding to the control angle μ of Fig. 4.32 and the corresponding flight path angle is in Fig. 4.33. Note the oscillations in γ during the trajectory control. The reentry trajectory is in Fig. 4.34 and the zoom in Fig. 4.35 shows the lateral turn realized by the capsule at the terminal phase of the reentry trajectory to achieve the target landing point.

Problems

4.1 Using the data in the table of Fig. 4.4, compare the maximum deceleration occurring in Earth and Mars atmosphere to the Soyuz capsule assuming zero aerodynamic efficiency.

4.2 Consider the Soyuz capsule with data shown in Fig. 4.4 and two reentry conditions: (i) $h_e = 100$ km, $V_e = 7900$ m/s, $\gamma_e = -1.5$ deg, (ii) $h_e = 100$ km, $V_e = 7900$ m/s, $\gamma_e = +0.5$ deg. Compare the values of the heat flux and of the total heat $Q = \int_{t_0}^{t_f} q(\tau)\, d\tau$.

---

[2] Parachute is open in the very last phase of the landing to further decrease the rate of descent.

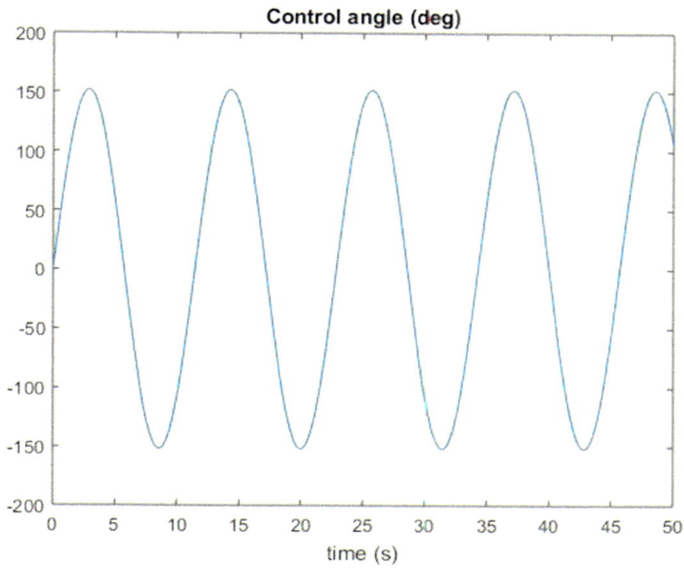

Fig. 4.32 The time history of the control variable μ

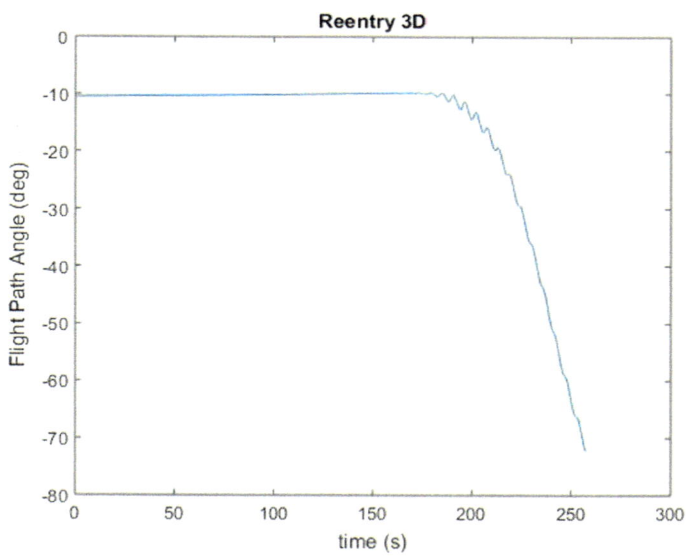

Fig. 4.33 The flight path angle during the controlled reentry trajectory

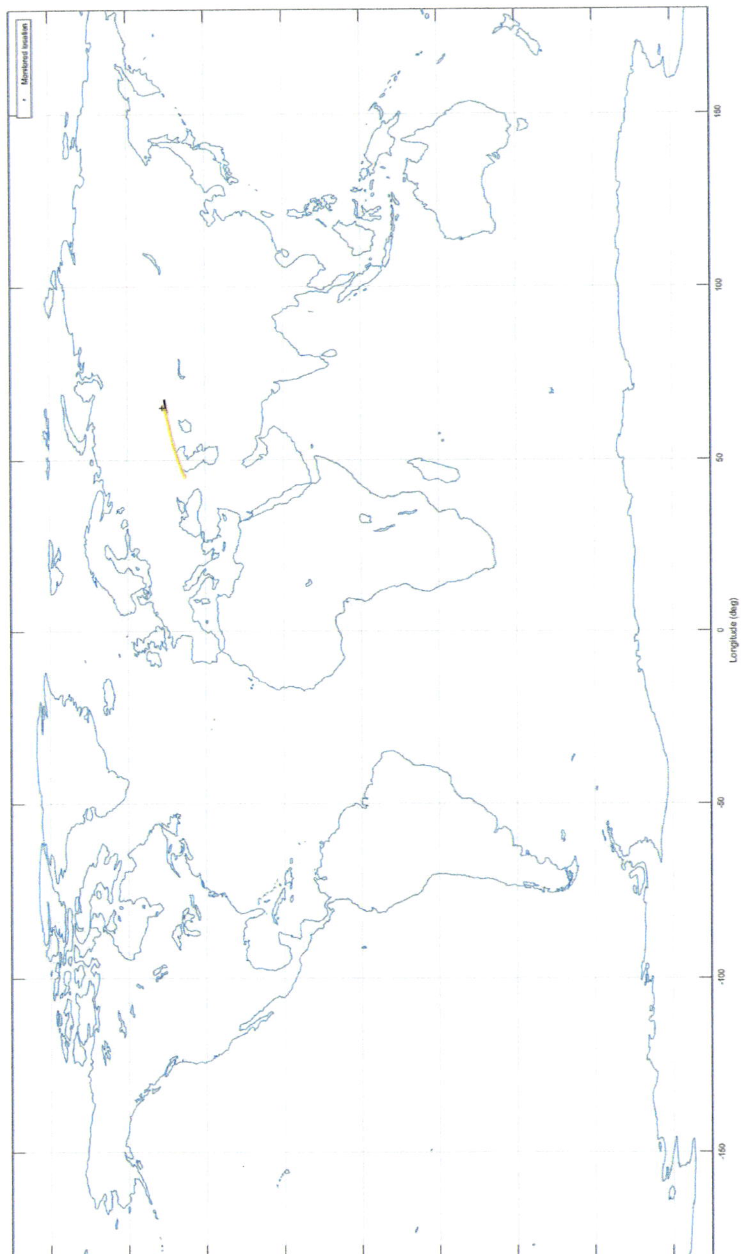

Fig. 4.34 The capsule ground track during reentry

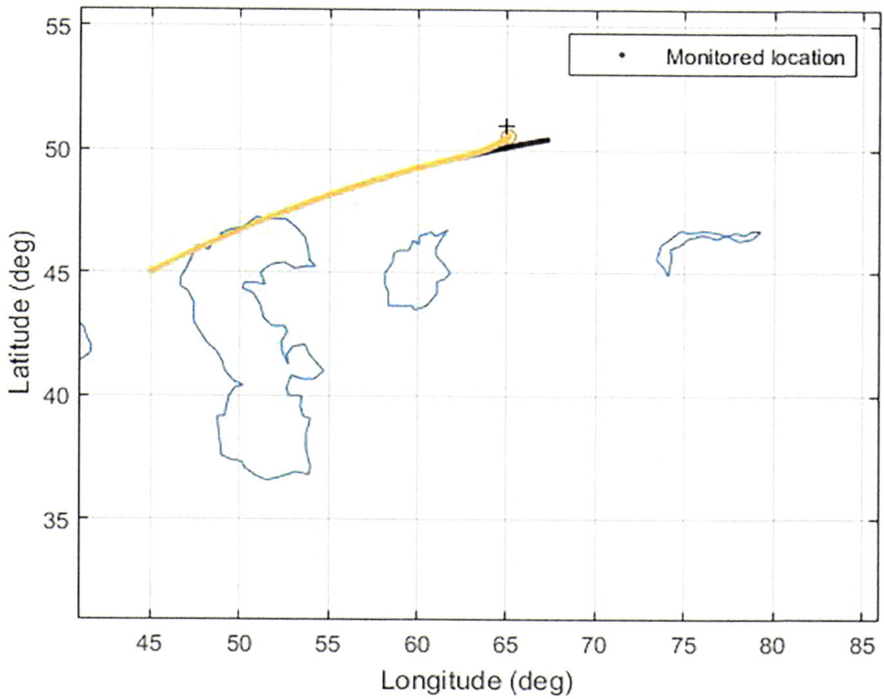

Fig. 4.35 A zoom of the capsule ground track

Note that the mass loss Δm of an heat shield with ablative material is proportional to the total heat: $\Delta m = \frac{Q}{H}$, with H effective heat of ablation [1].

Solutions

Problems of This Chapter

4.1 The Soyuz parameters and entry conditions are in the table of Fig. 4.4, in particular $V_e = 7.9\,\text{km/s}$, $\gamma_e = -1.5\,\text{deg}$.

The exponential model of the atmosphere with respect to the altitude is

$$\rho = \rho_0\, e^{-\beta\, h}$$

where

$$\rho_0 = 1.22\text{kg/m}^3 \; - \; for\ Earth\ , \;\; \rho_0 = 0.02\text{kg/m}^3 \; - \; for\ Mars$$

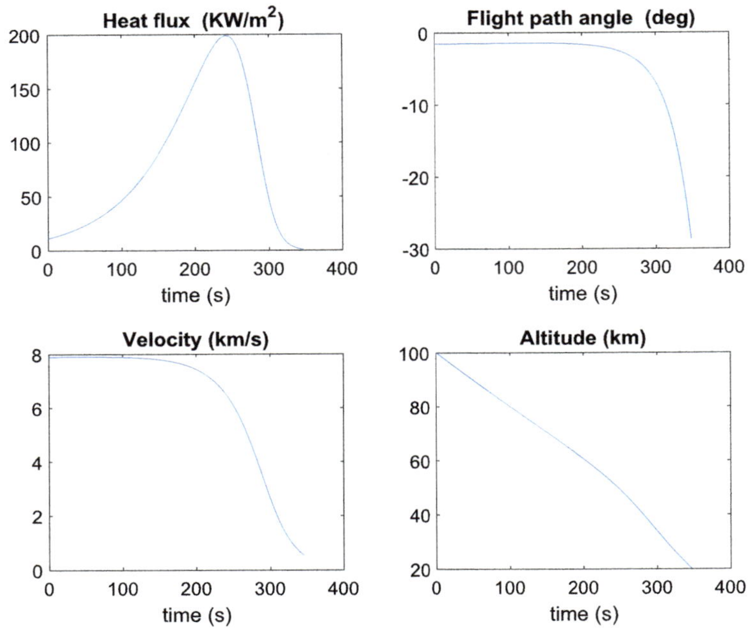

Fig. 4.36 Data of direct reentry of the Soyuz capsule

$$\beta = 0.1414\ 1/\text{km}\ -\ for\ Earth\ ,\ \beta = 0.091\ 1/\text{km}\ -\ for\ Mars$$

According to (4.16) the maximum deceleration is:

$$a_{max} = -4.33\,g\ -\ for\ Earth\ ,\ a_{max} = -2.79\,g\ -\ for\ Mars$$

occurring at the altitudes (4.14):

$$h^* = 46\,\text{km}\ -\ for\ Earth\ ,\ h^* = 31\,\text{km}\ -\ for\ Mars$$

Note however that the entry conditions at Mars may occur at rather higher velocity than V_e in direct interplanetary Earth to Mars trajectory.

4.2 Integration of the equation of the reentry Problems (4.1)–(4.4) with Soyuz parameters taken from the table of Fig. 4.4 and the standard reentry conditions (i) produces the data of Fig. 4.36. The integration is stopped when the altitude is equal to 20 km. A direct reentry is performed. The peak of heat flux is about 200 KW/m^2 and the total heat is $Q = 28400\,\text{KJ}/\text{m}^2$. The entry conditions (ii) with final altitude equal to 20 km are in Fig. 4.37. The reentry trajectory in not direct: two entries into the atmosphere are performed. The peak of heat occurs in the second entry and it is equal to 160 KW/m^2. The total heat accumulated in the two atmospheric entries is

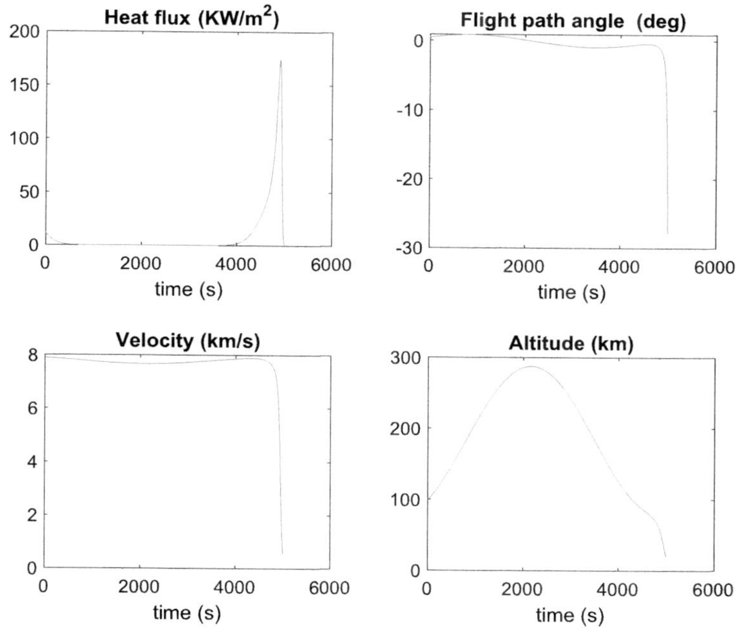

Fig. 4.37 Data of skip reentry of the Soyuz capsule

equal to $Q = 43285\ \mathrm{KJ/m^2}$. Skip reentry trajectories reduce the peak of heat while increasing the total heat.

References

1. Loh, W.: Re-entry and Planetary Entry, vol. 1. Springer, Berlin (1968)
2. Regan, F., Anandakirshnan.: Dynamics of the Atmospheric Reentry. AIAA Education Series (1993)
3. Hunt, G.: Apollo reentry guidance and navigation equations and flow logic, MSC Internal note 65-FM-157 (1965)
4. Bennet, M.: Reentry matters (2019). Available via AULIS https://www.aulis.com/re-entrymatters.htm
5. Hirschel, E.H, Weiland, C.: Selected Aerothermodynamic Design Problems of Hypersonic Flight Vehicles. Springer, Heidelberg (2009)
6. Johnson, G.W.: Blunt Capsules Drag Data, published on 19 August 2012 at exorocketman.blogspot.com
7. Miele, A.: Flight Mechanics. Addison-Wesley (1962)
8. Hoerner, S.F.: Fluid Dynamic Drag, self-published by the author (1965)
9. Allan, H., Egger, A.J.: A study of the motion and aerodynamic heating of ballistic missiles entering the earth's atmosphere at high supersonic speeds, NASA Technical Report 1381 (1958)
10. Vihn, N.X.: Optimal Trajectories in Atmospheric Flight. Elsevier, London (1981)
11. Cox, K.J.: Shuttle guidance, navigation and control design, Nasa Technical Memorandum TM-X-67709 (1971)
12. Chern, J., Vihn, N.X.: Optimal reentry trajectories of a lifting vehicle, Nasa Contractor Report 3236 (1980)

9783031757594